TEXTILE PROCESSES

TEXTILE PROCESSES

QUALITY CONTROL AND DESIGN OF EXPERIMENTS

FIRST EDITION

Professor D.Sc. GEORGI BORISOV DAMYANOV

Associate Professor PhD DIANA STOYANOVA GERMANOVA-KRASTEVA

Department of Textile Engineering
Technical University of Sofia

MOMENTUM PRESS

MOMENTUM PRESS, LLC, NEW YORK

Textile Processes: Quality Control and Design of Experiments

First published by Momentum Press®, LLC
222 East 46th Street, New York, NY 10017
www.momentumpress.net

ISBN-13: 978-1-60650-387-4 (hard back, case bound)
ISBN-10: 1-60650-387-1 (hard back, case bound)
ISBN-13: 978-1-60650-389-8 (e-book)
ISBN-10: 1-60650-389-8 (e-book)

DOI: 10.5643/9781606503898

Cover design by Jonathan Pennell
Interior design by Exeter Premedia Services Private Ltd., Chennai, India

10 9 8 7 6 5 4 3 2 1

Printed in the United States of America

Contents

Preface

This book has been written for textile experts, researchers, students, and PhD students in the field who work on the control and optimization of textile processes and the quality of the manufactured products.

The book presents a review of the methods for experiment design and for deriving and optimization of mathematical models. The individual models are illustrated by numerical examples, which allow for easier comprehension and implementation of the methods in practice.

The first part of the book is for those experts who are not familiar with the foundations of mathematical statistics. There, in a very accessible form, they will find a review of the main notions of mathematical statistics used in textile processes modeling. The review covers the main properties of the laws on distribution of random variables, the types of statistical estimates, and the ways to determine them, which are used most often in the textile practice.

There is a description of the algorithms for examination of various statistical hypotheses, as well as their graphic application for control of the technological process—the control charts.

There is a brief presentation of the correlation analysis and the analysis of variance (ANOVA), their significance, and mode of application.

The "Design of experiments" section gives a description of the particulars of mathematical modeling and preparation of experiments. There is a review of the capabilities and the main stages of the regression analysis.

The full and fractional factorial experiments for derivation of linear models are clarified with examples. There is a description of the simplex method and the Box–Wilson (experiment design) method for definition of the extreme area. There is a review of the three most wide-spread designs for derivation of second order mathematical models: the rotational central composite design, the orthogonal central composite design, and the optimal design.

You will find examples of the ways to present in a graphic manner the area, described by the model, and the ways to optimize the model. Explanations have also been provided on how to interpret the results of the designed experiment.

Special attention has been given to the specifics of the Taguchi method of experiment design.

We have strived to create a book which clarifies, in an accessible manner, the methods for experiment design and for optimization of the derived models. We hope that with the examples provided we will be useful to the experts and researchers in their efforts to increase the quality in the area of textiles.

Georgi Borisov Damyanov
Diana Germanova-Krasteva

About the Authors

Professor D.Sc Georgi Borisov Damyanov was born on 1927 in Sofia. He obtained a degree in Textile Engineering (1951) and Doctor of Science (1975) from Moscow Textile University and has specialized in textile technology in Poland, Russia, England, and the USA. Professor Damyanov started his professional career as a manager and chief engineer in various textile factories in Sofia (1951–1955).

From 1985 to 1997 he worked as member and president of the board of directors of textile companies. For many years he has been president of Bulgarian Association of Textile Engineers.

In 1956 he entered academia as a lecturer, became associate professor, and ultimately a full professor in 1969 in the Textile department of the Technical University of Sofia. He has lectured on textile technology at the University of Economics, Academy for Social Sciences and also in Poland, USA, and North Korea.

For many years Professor Damyanov has been head of the Textile department and also Dean and Prorector in the Technical University Sofia. His scientific interests include investigation of textile machinery and technology, application of radioactive isotopes in textile technology, and mathematical optimization of textile processes. Professor Damyanov is the author of over 250 research works published in professional textile journals in ten countries. As author and co-author has also written 41 textile textbooks and monographs, four of which are in English and Russian.

In 1966 Professor Damyanov was a visiting professor in the Textile department of the University of Manchester, England and in 1968–1969 he was a visiting professor at MIT in Boston, USA.

Associate Professor. PhD Diana Stoyanova Germanova-Krasteva was born in 1967 in Sofia. In 1991 she graduated from the Technical University of Sofia as a Master of Science in Textile Engineering. She began working as an editor in the scientific journal *Textile Industry* issued by the Scientific and Technical Union of Textile, Clothing and Leather.

In 1993 Diana Krasteva became professor in the Department of Textile Engineering. She leads classes in textile testing, textile materials, quality management, textile production, and others.

Her interests are also in the field of application of simulation modeling and statistical methods for modeling and optimization of textile processes.

In 2005 she obtained a PhD degree and since 2007 is an Associate Professor in the Department of Textile Engineering at the Technical University of Sofia.

Professor Krasteva specializes in Belgium, Spain, and Germany. She is the author of over 50 publications in Bulgarian, German, Swiss, English, Polish, and Romanian scientific journals. She has published three textbooks in the field of textiles, textile testing, and statistical analysis.

List of Figures

List of Tables

PART I

INTRODUCTION TO MATHEMATICAL STATISTICS

The production of high quality textiles is based on the excellent knowledge of the properties of textile materials. The establishing of these properties is achieved by testing. The test results are data, the processing of which is performed according to the relationships of the theory of probability and mathematical statistics.

The mathematical foundations of statistics were laid during the seventeenth century with the development of probability theory by Blaise Pascal and Pierre de Fermat. The extensive use of computers nowadays allows the mass application of statistical calculations on large data volumes and has lead to the development of new computational methods.

In Part I, the main concepts and methods of processing and analyzing the results obtained, and applying them in the textile processes evaluation will be presented.

I.1. General Terms and Definitions

Mathematical statistics views the regularities and methods used for the study of mass phenomena that are characterized by individuals that are homogenous in a qualitative aspect. The combination of a large number of individuals (cases) is hereby referred to as *statistical population.*

The statistical population can be:

- *General*, which includes all cases of the studied mass phenomenon, or
- *Representational* (*sample*), which includes part of the cases of the general population, the characteristics of which can be used to draw conclusions regarding the characteristics of the general population.

The volume of the general population (the batch) is usually assigned with N and the sample with n.

Mathematical statistics is based on *probability theory* that uses a series of specific terms such as test, event, random variable, probability, and so forth.

- A *test* (*observation*) is the implementation of a specific sum of conditions.
- An *event* is any fact that has occurred as a result of holding the test.
- *Probability of occurrence of a specific event* is a number that expresses the possibility of occurrence of the specific event.
- A *valid event* is an event which always occurs as a result in the test. The probability of occurrence of this event is 1 or 100%.
- An *impossible event* is an event which never occurs as a result in the test. The probability of occurrence of this event is 0.
- A *random event* is an event which could or could not occur as a result of the test. The probability of its occurrence ranges between 0 and 1, or between 0 and 100%.
- A *practically impossible event* is the event for which the probability of occurrence is very close to 0, for example 0.1, 0.05, 0.01, 0.001.
- A *practically valid event* is an event for which the probability of occurrence is very close to 1, for example 0.9, 0.95, 0.99, 0.999.
- A *significant level* is the probability of occurrence for a practically impossible event α.
- The *confidence level* is the probability $\gamma = 1 - \alpha$ for the occurrence of a practically valid event. In the textile production sector if no special instructions have been given, operations will be performed at the significant level $\alpha = 0.05$.
- A *random variable* is a variable, which as a result of the performance of the test can adopt different values which could not be predicted in advance.

A random variable can be discrete or continuous. A *discrete random variable* can adopt only separate values isolated from one another, for example, the number of defective articles in a single batch, the number of fibers in a band, the number of stops of the machine. A *continuous random variable* can be adopted by any value in a given finite or infinite interval, for example, yarn strength.

A *random variables distribution* expresses the dependence between the possible values of the random variable and their respective probabilities. Random variables distribution can be described by means of the distribution function, or the density of probability distribution, or the order of distribution.

Distribution function $F(x)$, also referred to as integral distribution function or integral distribution principle, expresses the probability of the random variable X to remain a smaller number than the number x:

$$F(x) = P\{X < x\}. \tag{I.1.1}$$

Figure I.1.1 presents the function of distribution of a continuous variable. The figure indicates that in this specific case, the probability $P\{X < x = 0\} = 0.5$.

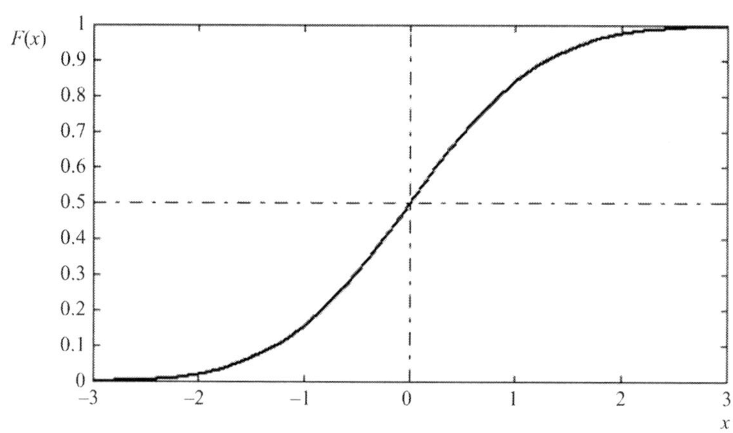

Figure I.1.1. Function of distribution of a continuous variable.

In case the variable is discrete, the distribution function $F(x)$ has a terraced form (Figure I.1.2). In case of Figure I.1.2, the following probability has been determined: $P\{X < x = 4\} = 0.265$.

The *density of probability distribution* $f(x)$, also referred to as density of probabilities, differential principle of distribution, or differential distribution function, is determined as a derivative of $F(x)$ along x:

$$f(x) = \frac{dF(x)}{dx}. \tag{I.1.2}$$

Distribution density exists only for continuous variables and is of the type shown in Figure I.1.3.

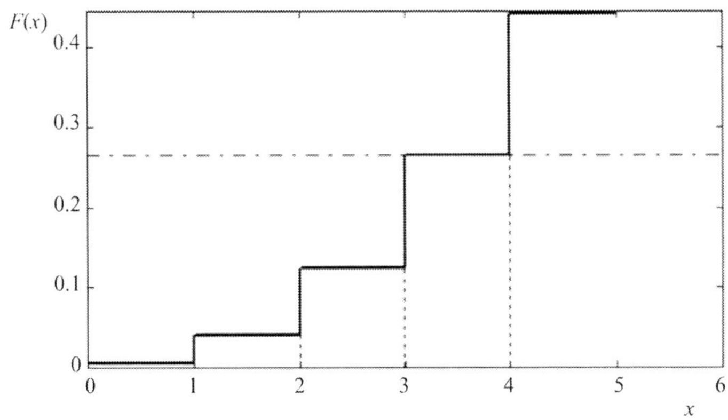

Figure I.1.2. Function of distribution of a discrete variable.

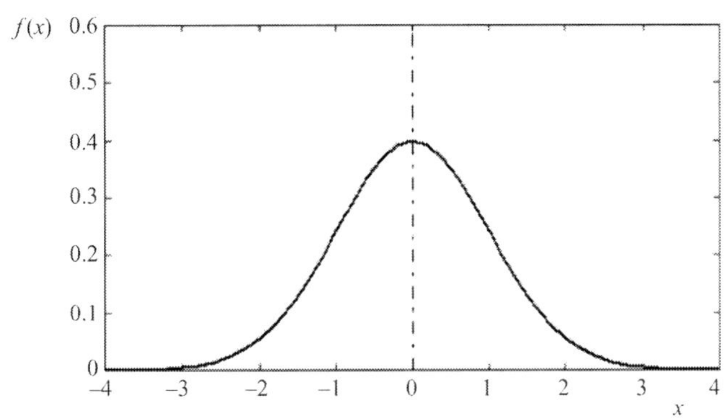

Figure I.1.3. Probability distribution function.

The *order of distribution* consists of the possible values of the discrete random variable and the probability for it to have the respective values. The order of distribution can be represented as a table or a graphic (Figure I.1.4) and the separate discrete values are not connected.

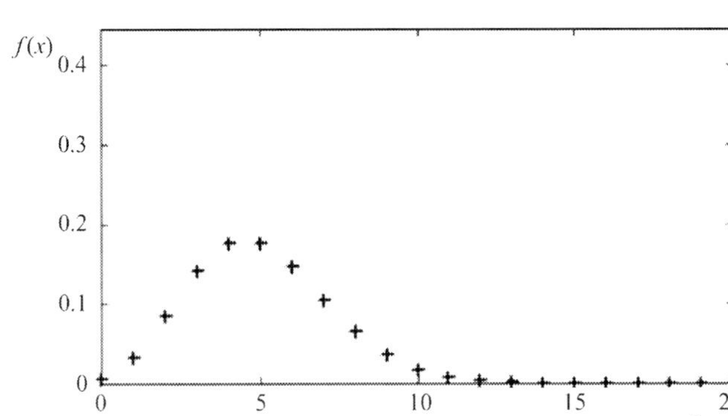

Figure I.1.4. Order of distribution.

PARAMETERS AND NUMERICAL CHARACTERISTICS OF THE RANDOM VARIABLE

Very frequently, instead of the distribution rule for characterization of some essential aspects of the random variable, separate numerical parameters and characteristics are used. The *distribution parameters* are variables which are directly involved in the distribution function or the probability distribution function. *Numerical characteristics* are characteristics expressing certain properties of the random variable distribution rule. The parameters (characteristics) can be divided into several groups: those referring to the location, the dispersion, and the moments.

CHARACTERISTICS FOR LOCATION

1. *Quantile* x_p—the value x_p of the random variable X, for which the probability

$$P\{X < x_p\} = p, \quad 0 < p < 1. \tag{I.1.3}$$

For example, the quantiles $x_{1/4}$, $x_{1/2}$, and $x_{3/4}$ separate the interval of the random variable X to four subintervals with equal possibilities for X being part of them (Figure I.1.5).

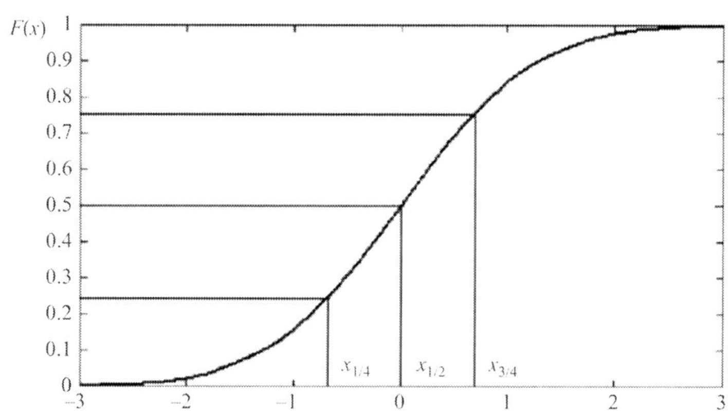

Figure I.1.5. Quantiles of the distribution function.

2. *Mathematical expectation* $M\{X\}$, μ_x, m_x—the average value of the random variable defined by the weight aspect according to the probability for its occurrence. For discrete random variables it is defined as

$$M\{X\} = \mu_x = m_x = \sum_i x_i p(x_i) \tag{I.1.4}$$

and for continuous variables as

$$M\{X\} = \mu_x = m_x = \int_{-\infty}^{\infty} x \cdot f(x) dx. \tag{I.1.5}$$

For certain types of distributions a mathematical expectation may not exist.

3. *Median* M_e—the point separating the interval of variation of the random variable X in two parts so that the probability of occurring therein would be ½, that is, the median is a quantile at $p = ½$.
4. *Mode* M_o—the value(s) of the random variable X, which in case of a continuous random variable is compliant with the local maximum of the probability density, and in case of a discrete random variable is compliant with the local maximum of the series of probabilities. Depending on the number of modes, the distributions can be single-mode or multimode. It is possible that a mode does not exist for a specific type of distribution.

   A single continuous distribution is symmetrical if

$$F(M_e - x) = 1 - F(M_e + x), \tag{I.1.6}$$

respectively,

$$f(M_e - x) = f(M_e + x). \tag{I.1.7}$$

For those distributions, the mathematical expectation matches the median and the mode.

DISPERSION CHARACTERISTICS

1. *Variance* $D\{X\}$, σ_x^2—the characteristic of dispersion of the random variable X around its mathematical expectation. For discrete random variables the following formula has been determined:

$$D\{X\} = \sigma_x^2 = \sum_i (x_i - \mu_x)^2 p(x_i), \tag{I.1.8}$$

and for continuous ones it is

$$D\{X\} = \sigma_x^2 = \int_{-\infty}^{\infty} (x - \mu_x)^2 f(x)dx. \tag{I.1.9}$$

The equivalent formula is frequently used as well:

$$D\{X\} = \sigma_x^2 = M\left\{(X - \mu_x)^2\right\} = M\left\{X^2\right\} - \left(M\{X\}\right)^2. \tag{I.1.10}$$

Figure I.1.6 represents the distribution density for three random variables with matching mathematical expectations $\mu = 0$ and variances, respectively, (a) $\sigma_1^2 = 0.25$, (b) $\sigma_2^2 = 1$, and (c) $\sigma_3^2 = 4$.

2. *Standard deviation* σ_x—characterizes the dispersion of the random variable X around its mathematical expectation and has the same dimension as X:

$$\sigma_x = \sqrt{D\{X\}}. \tag{I.1.11}$$

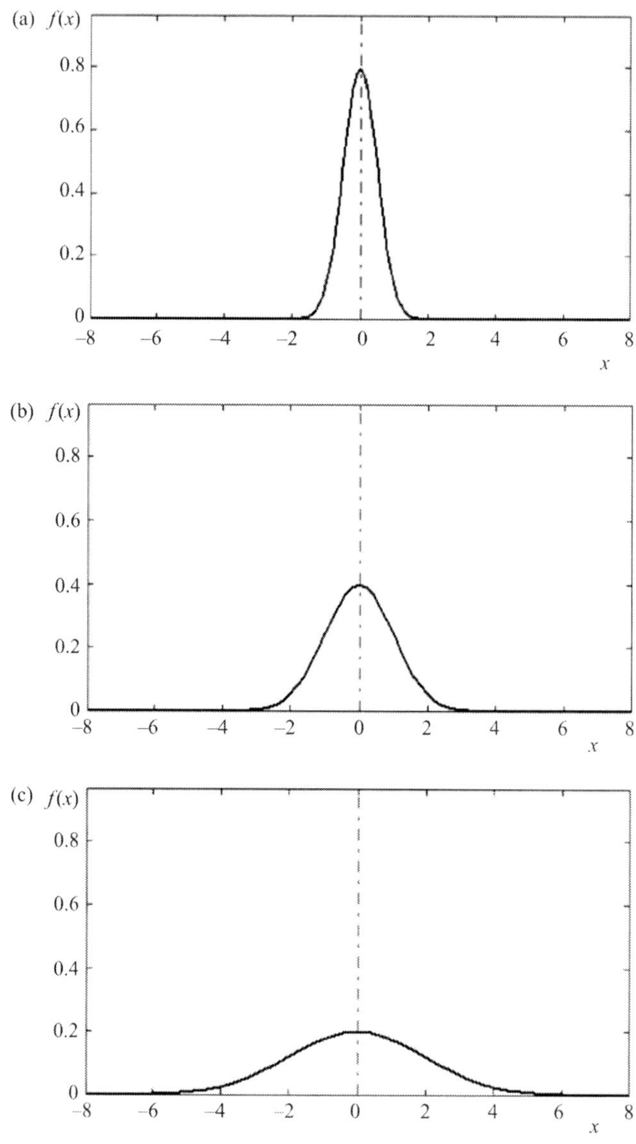

Figure I.1.6. Probability distribution functions with different variances.

3. *Coefficient of variation*—a relative characteristic for evaluation of dispersion of the random variable. Percentage-wise it is determined according to the following formula:

$$v = \frac{\sigma_x}{\mu_x} \cdot 100. \tag{I.1.12}$$

4. *Linear deviation d*—characterizes the average absolute deviation of the random variable:

$$d\{X\} = M\{|X - \mu_x|\}. \tag{I.1.13}$$

5. *Range R*—the difference between the maximum and the minimum values of X, for which $f(x) \neq 0$. It is defined as

$$R = X_{max} - X_{min}. \tag{I.1.14}$$

MOMENTS

1. *Initial moments* a_k—the initial moment from rank k is the average numerical characteristic:

$$a_k = M\left\{X^k\right\}. \tag{I.1.15}$$

The initial moment of rank 0 always equals 1:

$$a_0 = M\left\{X^0\right\} = 1, \tag{I.1.16}$$

and the initial first rank moment always equals the mathematical expectation:

$$a_1 = M\left\{X^1\right\} = \mu_x. \tag{I.1.17}$$

2. *Central moments* μ_k—the central moment from the kth rank is determined analogically to the initial rank and the random variable is centered according to the mathematical expectation:

$$\mu_k = M\left\{\left(X - \mu_x\right)^k\right\}. \tag{I.1.18}$$

The central moment from first rank ($k = 1$) always equals 0:

$$\mu_1 = M\left\{\left(X - \mu_x\right)^1\right\} = 0 \tag{I.1.19}$$

and the central moment from second rank ($k = 2$) always equals the variance:

$$\mu_2 = M\left\{\left(X - \mu_x\right)^2\right\} = \sigma_x^2. \tag{I.1.20}$$

The central moments of uneven rank of random variables with symmetric distributions equals 0.

3. *Asymmetry (skewness)* S_k—characterizes the degree of asymmetry of a distribution around its mean. It is determined by the relation:

$$S_k = \frac{\mu_3}{\sigma_x^3}. \tag{I.1.21}$$

The skewness value is 0 with symmetrical distributions ($\mu_3 = 0$), its value being positive when the mode is situated before the median and negative when the mode is situated after the median (Figure I.1.7).

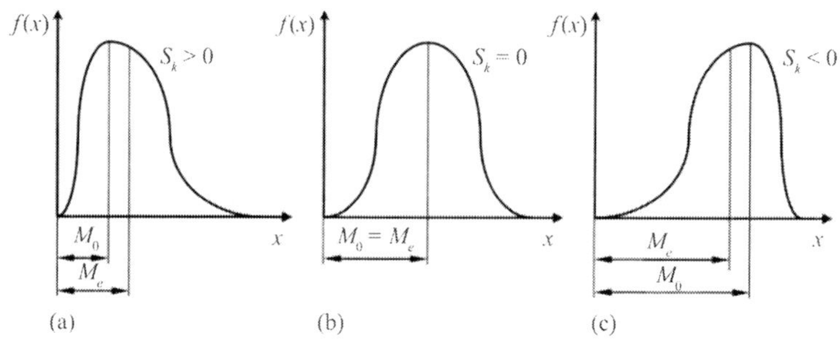

Figure I.1.7. Distributions with positive asymmetry, symmetrical distribution, and negative asymmetry.

4. *Kurtosis (declivity) E*—characterizes the relative peakness or flatness of a distribution in comparison with the normal distribution. It is determined according to the following formula:

$$E = \frac{\mu_4}{\sigma_x^4} - 3. \tag{I.1.22}$$

Positive kurtosis indicates a relatively peaked distribution. Negative kurtosis indicates a relatively flat distribution (Figure I.1.8).

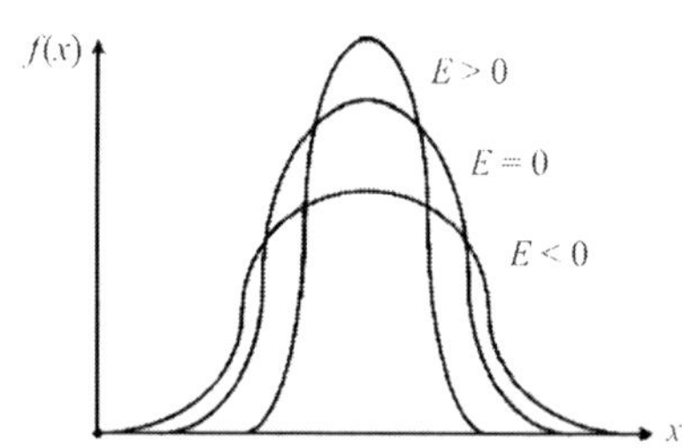

Figure I.1.8. Distribution with positive, zero, and negative kurtosis.

The kurtosis is 0 when the form of the distribution curve is such that the following is true:

$$\mu_4 = 3\sigma_x^4. \tag{I.1.23}$$

5. *Correlation moment (co-variation factor)* μ_{xy}—the central moment characterizing the degree of linear dependency of two random variables X and Y:

$$\mu_{xy} = M\left\{\left(X - \mu_x\right)\left(Y - \mu_y\right)\right\}. \tag{I.1.24}$$

6. *Linear correlation factor* ρ_{xy}—the standardized correlation moment, the values of which fall within the margins [–1; 1]. Standardization is completed through the division of the correlation moment into the product of the standard deviations of the two random variables σ_x and σ_y:

$$\rho_{xy} = \frac{\mu_{xy}}{\sigma_x \cdot \sigma_y}. \tag{I.1.25}$$

For $|\rho_{xy}| = 1$ between the two random variables, a linear functional dependency exists. When $-1 < \rho_{xy} < 1$ between the variables x and y, a statistical (stochastic) dependency exists, whereas the closer the value of $|\rho_{xy}|$ is to 0, the stronger the representation of this value is. If ρ_{xy} is positive, the increase in values of x leads to increase in values of y. If ρ_{xy} is negative, the increase of x results in diminishing y and a reverse dependency will be discussed. When $\rho_{xy} = 0$, the random variables are uncorrelated. If the random variables x and y are independent, they are noncorrelated. The reverse is correct only for principles of normal distribution. All separate cases have been presented in Figure I.1.9 [(a) reverse statistical dependency, (b) no correlation, (c) straight statistical dependency].

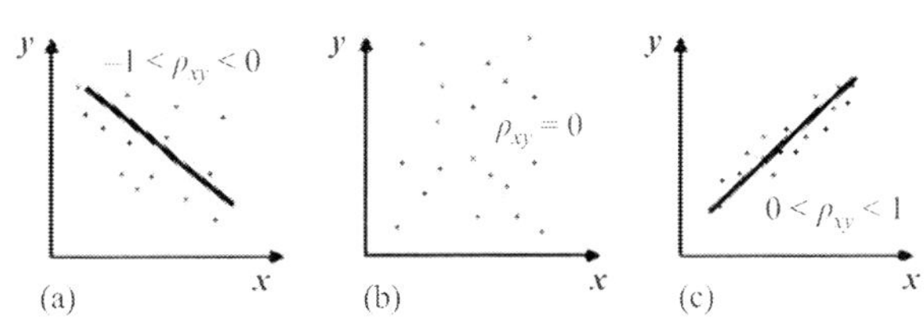

Figure I.1.9. Different cases of correlation dependencies.

PROPERTIES OF NUMERICAL CHARACTERISTICS

Textile practice often requires working with variables that are a sum or a product of two or more random variables, as well as a product of a random variable and a constant. In such cases, the following properties of the numerical characteristics can be used.

1. The mathematical expectation of the constant ($C = const$) is equal to the variable itself:

$$M\{C\} = C = const. \tag{I.1.26}$$

2. The mathematical expectation of the sum of random variables equals the sum of their mathematical expectations:

$$M\left\{\sum_{i=1}^{n} X_i\right\} = \sum_{i=1}^{n} M\{X_i\}. \tag{I.1.27}$$

The property is also valid for subtraction of random variables. In this case, a minus sign will be placed before the respective random variables.

3. The mathematical expectation of the product of two random variables equals the sum of the product of the mathematical expectations of the two components and the correlation moment:

$$M\{X \cdot Y\} = M\{X\} \cdot M\{Y\} + \mu_{xy}. \tag{I.1.28}$$

If the random variables are noncorrelated:

$$M\{X \cdot Y\} = M\{X\} \cdot M\{Y\} = \mu_x + \mu_y. \tag{I.1.29}$$

4. The mathematical expectation of the product of a random variable with a constant follows:

$$M\{C \cdot X\} = C \cdot M\{X\}. \tag{I.1.30}$$

5. The variance of a constant is zero:

$$D\{C\} = 0. \tag{I.1.31}$$

6. The variance of the product of a random variable and a constant is

$$D\{C \cdot X\} = C^2 \cdot D\{X\}. \tag{I.1.32}$$

7. The variance of a sum or a margin of two random variables equals

$$D\{X \pm Y\} = D\{X\} + D\{Y\} \pm 2\mu_{xy}. \tag{I.1.33}$$

Result: If the random variable $X = \pm X_1 \pm X_2 \pm \cdots \pm X_n$ and each pair of random variables are mutually uncorrelated, then

$$D\{X\} = D\{X_1\} + D\{X_2\} + \cdots D\{X_n\}. \tag{I.1.34}$$

I.2. LAWS OF RANDOM VARIABLES DISTRIBUTION

This section presents those probability distributions that are used most often in the textile practice.

CONTINUOUS DISTRIBUTIONS

1. *Normal (Gaussian) distribution*

The distribution density law is as follows:

$$P(x) = \frac{1}{\sigma\sqrt{2\pi}} e^{-\frac{(x-m_x)^2}{2\sigma^2}}. \quad \text{(I.2.1)}$$

The curve of the distribution is bell-shaped and with the increase in the absolute value of X ($|x| \to 0$), it tends asymptotically to the x-axis. The distribution, respectively the form of its graph, is defined by two parameters. The mathematical expectation m_x defines the location of the curve (Figure I.2.1) and the standard deviation σ defines the variance around it. When the standard deviation is big the curve is flat, and when it is small the curve is convex (Figure I.1.6).

The normal distribution is typical for many textile properties and is used when the property that is being studied is dispersed symmetrically in relation to the central value, for example, strength, count, elongation, linear dimensions (length, width, diameter), and so on. Very often, when studying random variables, including those distributed according to the normal distribution law, the values are centered and standardized. A *centered random variable* is a variable the values of which lie symmetrically around zero. The centering is done by extracting the respective mathematical expectation from the individual variable values:

$$\overset{o}{X} = X - m_x. \quad \text{(I.2.2)}$$

The *standardization of a random variable* is done by dividing its centered value by the standard deviation:

$$X_N = \frac{X - m_x}{\sigma} = \frac{\overset{o}{X}}{\sigma}. \tag{I.2.3}$$

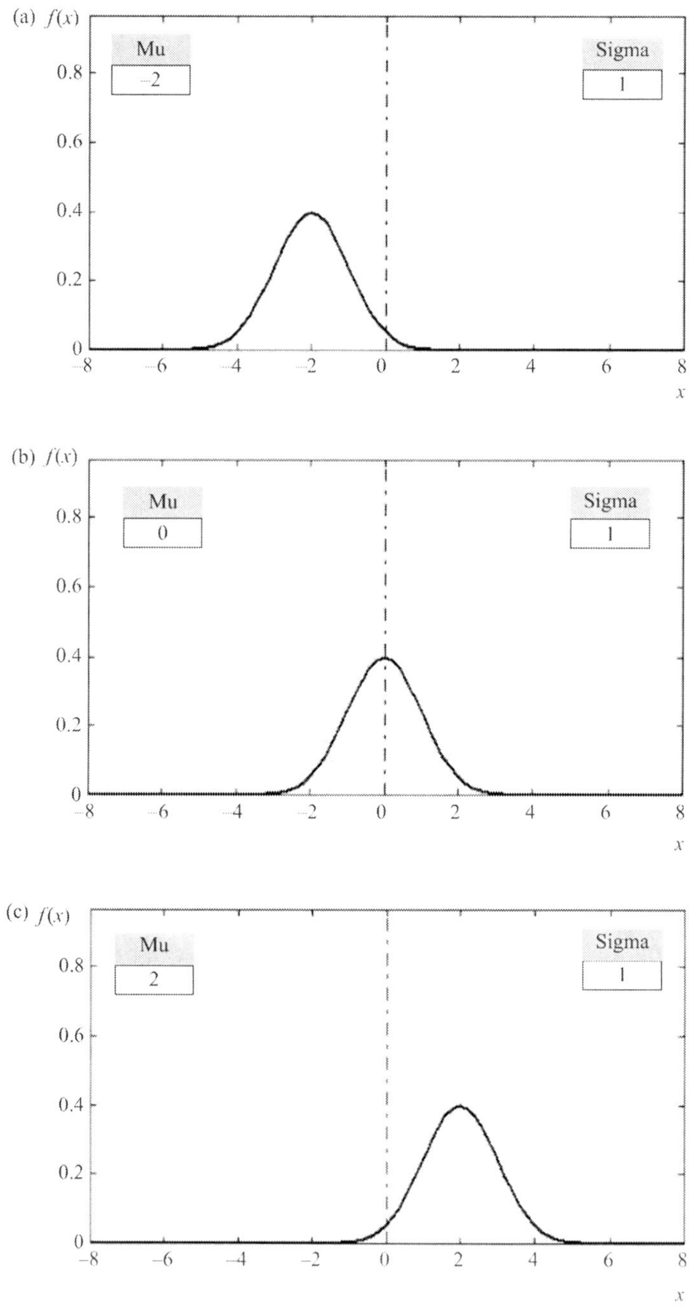

Figure I.2.1. Movement of the normal distribution in conformity with the mathematical expectation.

A standardized random variable has a mathematical expectation of 0 and a variance of 1—$N(0; 1)$.

For example, when determining the linear density of linear products, instead of putting down the value of the separate sections, it is better to give the deviations from the mean value in percentage. With the help of the standardized random function it is also very easy to determine the share of the values in some typical intervals (Figure I.2.2):

$$(m_x \pm \sigma) \rightarrow 68.26\%$$

$$(m_x \pm 2\sigma) \rightarrow 95.46\%$$

$$(m_x \pm 2.575\sigma) \rightarrow 99\%$$

$$(m_x \pm 3\sigma) \rightarrow 99.73\%.$$

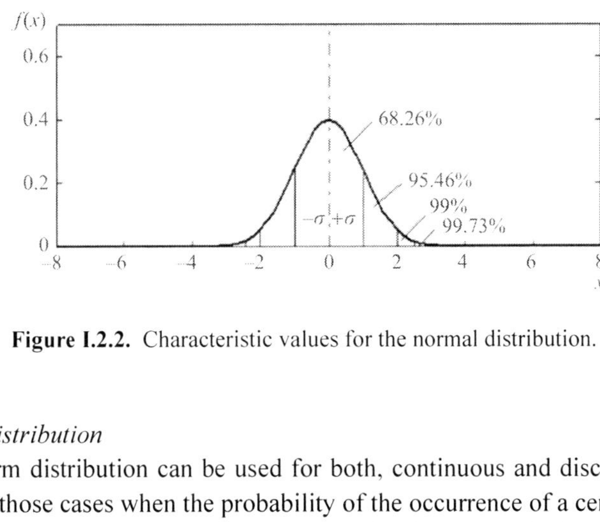

Figure I.2.2. Characteristic values for the normal distribution.

2. *Uniform distribution*

The uniform distribution can be used for both, continuous and discrete variables. It is applied in those cases when the probability of the occurrence of a certain value within a particular interval $[a; b]$ is the same. The distribution density is described with the help of the following law:

$$P(x) = \begin{cases} \dfrac{1}{b-a}, & a \le x \le b \\ 0, & b \le x \le a \end{cases}. \tag{I.2.4}$$

The mathematical expectation is defined as follows:

$$m_x = \frac{a+b}{2}, \tag{I.2.5}$$

and the formula for the variance is

$$\sigma^2 = \frac{(b-a)^2}{12}. \tag{I.2.6}$$

The graph of the uniform distribution is shown in Figure I.2.3.

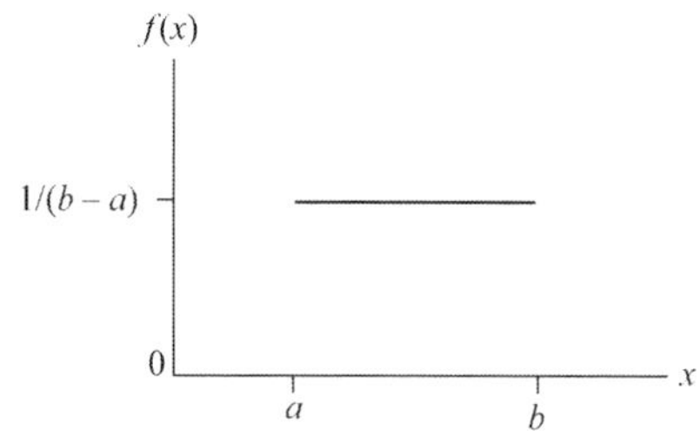

Figure I.2.3. Uniform distribution.

DISCRETE DISTRIBUTIONS

1. *Binomial distribution*
   The binomial distribution is the most widespread discrete distribution. It shows the probability of a certain event to occur x times out of n cases (repetitions). The distribution law is defined as follows:

$$P(x) = c_n^x p^x q^{n-x} = \frac{n!}{x!(n-x)!} p^x q^{n-x}, \tag{I.2.7}$$

where p is the probability of occurrence of a certain event and
$q = 1 - p$ is the probability of nonoccurrence of the particular event.

The mathematical expectation is

$$m_x = n \cdot p \tag{I.2.8}$$

and the variance is

$$\sigma^2 = n \cdot p \cdot q. \tag{I.2.9}$$

Figure I.2.4 presents different cases of binomial distribution when there is a change in the number of tests n and the probability of occurrence of the event p. Cases (a) and (b) show the distribution when there is a change in the number of tests from 5 to 10, the probability of occurrence being $p = 0.1$. The probability of nonoccurrence is $q = 0.9$.

The distribution is asymmetrical. In cases (c) and (d) the number of tests is changed again, but the probability of occurrence is now $p = 0.5$, respectively, and the probability of nonoccurrence is $q = 0.5$. The distribution is symmetrical. It can be seen that in case of a symmetrical distribution and increase in the number of tests, the distributions tends to normal.

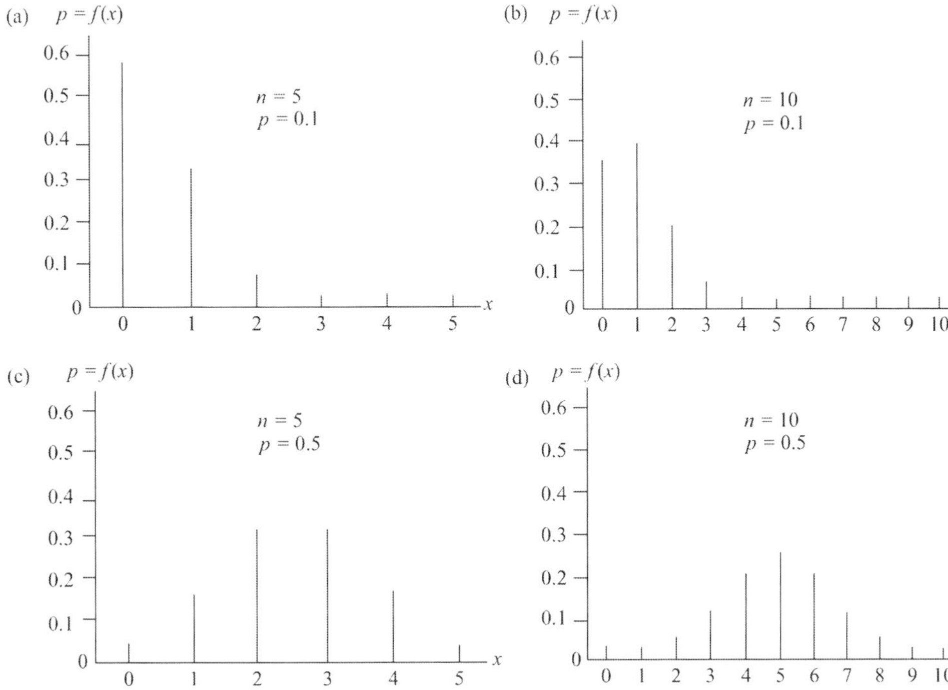

Figure I.2.4. Binomial distributions.

In order to illustrate the binomial distribution, the most common example used is the example of determining the probability when rolling a die. For example, let us try to determine the probability of rolling a three twice in 10 throws. The probability is determined when the following values are substituted in formula (I.2.7): number of tries (throws) $n = 10$, probability of the event "rolling a three" $p = \frac{1}{6}$ (the possible options are 6, each has an equal chance of occurrence), and the desired number of occurrences of the event is $x = 2$. After substituting, the probability of rolling a three twice in 10 throws can be determined:

$$P(x) = \frac{10!}{2!(10-2)!}\left(\frac{1}{6}\right)^2\left(\frac{5}{6}\right)^{10-2} = 0.29.$$

The binomial distribution is often used in the textile production when qualifying the production.

Example

In the production of hemp technical fabrics for sacks it has been established that out of 100 samples of the fabric an average of 50 have defects. Determine the probability of having a maximum of two samples with defects out of a total of three samples.

First determine the probability of having a fabric with a defect:

$$p = \frac{50}{100} = 0.5.$$

In order to satisfy the condition of having a maximum of two defective samples, the possible options are no defective sample, one defective sample, and two defective samples. The probability of having no defective samples when the volume of the sample is $n = 3$ is

$$P(0) = \frac{3!}{0!(3-0)!}(0.5)^0(0.5)^{3-0} = 0.125.$$

The probability to have one defective sample is

$$P(1) = \frac{3!}{1!(3-1)!}(0.5)^1(0.5)^{3-1} = 0.375.$$

The probability to have two defective samples is

$$P(2) = \frac{3!}{2!(3-2)!}(0.5)^2(0.5)^{3-2} = 0.375.$$

Therefore, the probability to have a maximum of two defective samples is

$$P(x \leq 2) = P(0) + P(1) + P(2) = 0.875.$$

2. *Poisson distribution*
   The Poisson distribution is an instance of the binomial distribution in which with the increase in the number of tests n the probability for the occurrence of the event p is decreased, so that $p \cdot n = \text{const} = \lambda$. Therefore, it is often said that the Poisson distribution is a distribution of the rare events. The distribution law is the following:

$$P(x) = \frac{e^{-\lambda}\lambda^x}{x!}, \quad x = 0, 1, 2, \ldots \tag{I.2.10}$$

What is typical about the Poisson distribution is that it is monoparametrical. The mathematical expectation and the variance equal the parameter λ:

$$m_x = \sigma^2 = \lambda. \tag{I.2.11}$$

Figure I.2.5 represents different cases of Poisson distribution when there is a change in the number of the parameter λ.

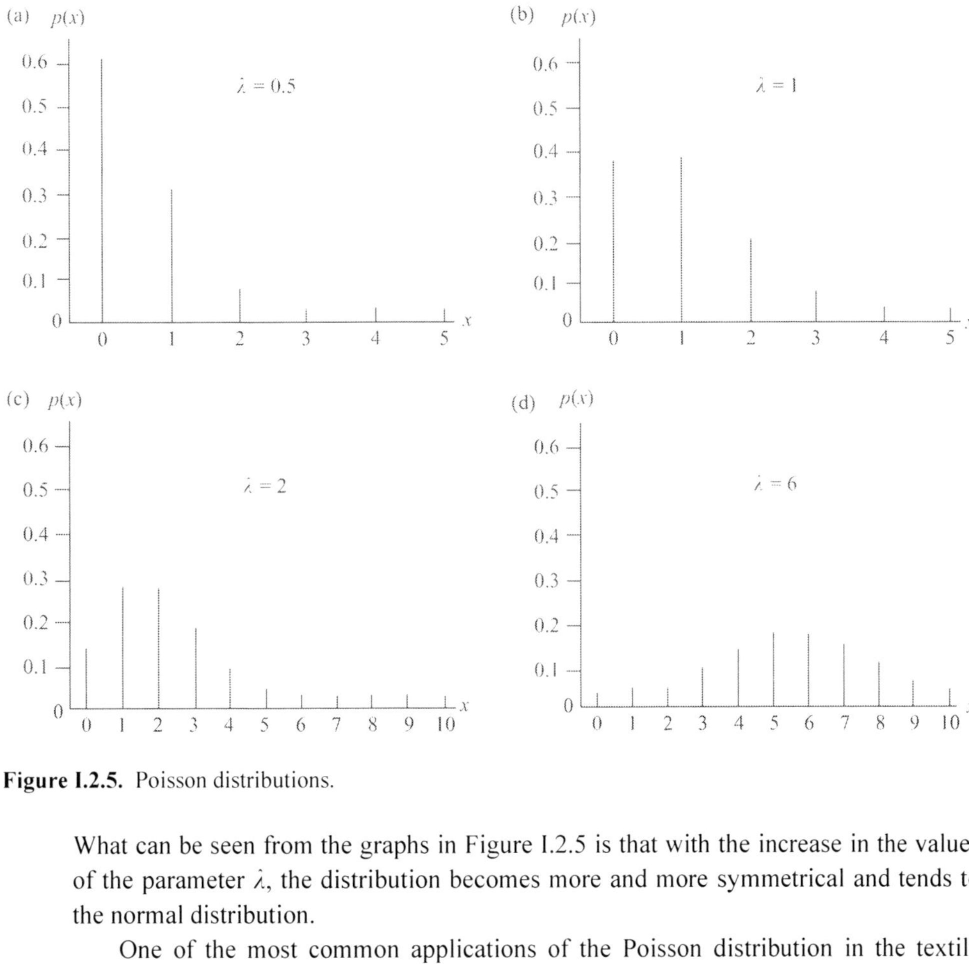

Figure I.2.5. Poisson distributions.

What can be seen from the graphs in Figure I.2.5 is that with the increase in the values of the parameter λ, the distribution becomes more and more symmetrical and tends to the normal distribution.

One of the most common applications of the Poisson distribution in the textile production is in determining the number of fibers in a section of a textile mill product. The irregularity is determined with the help of the coefficient of variation according to the formula given (see formula (I.1.12)). Taking into account that $m_x = \overline{n}$, where $\overline{n}$ is the mean number of fibers in the product section and the property of the Poisson distribution for equality between the variance and the mathematical expectation (see formula I.2.10), the coefficient of variation is the following:

$$v = \frac{\sigma}{m_x}100 = \frac{\sqrt{\overline{n}}}{\overline{n}}100 = \frac{100}{\sqrt{n}}. \qquad \text{(I.2.12)}$$

The resulting value of the irregularity (coefficient of variation) is theoretic as it has been obtained on the basis of the random distribution of fibers without taking into account the irregularity of the fibers and the impact of machine parts. The real irregularity of the products is much higher.

The Poisson distribution is also used very often in assessing the breakage in processing.

Example

What is studied is the breakage of a spinning machine with 96 spindles. It has been established that there are 80 breakages for 2 h. What has to be found out is the number of spindles for which the number of breakages is expected to be 0, 1, 2, 3, 4, and so on.

The mean number of breakages for one spindle for a period of 2 h is

$$\bar{x} = \frac{80}{96} = 0.833.$$

The probability that there will be no breakage within a period of 2 h is

$$P(0) = \frac{e^{-0.833} 0.833^0}{0!} = 0.435.$$

The probability of one breakage is: $P(1) = \frac{e^{-0.833} 0.833^1}{1!} = 0.363.$

The probability of two breakages is: $P(2) = \frac{e^{-0.833} 0.833^2}{2!} = 0.151.$

For three breakages: $P(3) = \frac{e^{-0.833} 0.833^3}{3!} = 0.047.$

For four breakages: $P(4) = \frac{e^{-0.833} 0.833^4}{4!} = 0.0087.$

In order to determine the number of spindles that will have the respective number of breakages, the probabilities are multiplied by the total number of spindles:

For a period of 2 h there will be no breakage on $96 \times 0.435 = 42$ spindles.

One breakage is expected on $96 \times 0.363 = 35$ spindles.
Two breakages on $96 \times 0.151 = 14$ spindles.
Three breakages on $96 \times 0.047 = 4$ spindles.
Four breakages on $96 \times 0.0087 = 1$ spindle.

There are other types of distributions that are used in textile production such as the Weibull and Gamma distributions, but the ones presented above are the most widely used.

I.3. STATISTICAL ESTIMATES

The definition of the numerical characteristics of the law, or respectively of the distribution function of a population of observed data, is called *statistical estimation* and the resulting numerical values are *statistical estimates.* The statistical estimates define the real values of the respective numerical characteristics only in an approximate manner. The reason is that they are calculated on the basis of a random sample with limited volume. Increasing the volume of the sample increases the accuracy of the estimates, which in turn is related to an increase in the test costs. One of the tasks of mathematical statistics is to find the minimum sample volume in order to obtain such a degree of accuracy in estimation that satisfies the needs of the researcher.

Mathematical statistics also makes use of different functions, called *statistics*, which depend solely on the observation data. Each statistic used for the estimation of a certain parameter or numerical characteristic of the distribution of the population is distributed according to some law around the respective parameter or characteristic. Under an analogy with the probability theory where the task is to find the most accurate value of a parameter or a numerical characteristic, this estimation is called point estimation, and the estimate—a *point estimate.* When with the help of the distribution of the statistic used is defined the interval in which there is a sufficient probability to find the estimated parameter or numerical characteristic, the estimation is called an interval estimation and the estimates—*interval estimates*.

CONDUCTING THE TEST

As a result of conducting the test, quantitative or qualitative estimates of the investigated indications are obtained, which are the respective *realization of the variable* when performing the test. The resulting data, recorded according to the order of their appearance, is called *primary range*.

The initial form of results recording does not allow a definition of the rule in the change of the indication. In order to provide a possibility for scientific research, first the data should be arranged according to value. What is obtained is the so-called *arranged range.* The arrangement can be in ascending $x_1 \leq x_2 \leq \cdots \leq x_i \leq \cdots \leq x_n$ or descending order $x_1 \geq x_2 \geq \cdots \geq x_i \geq \cdots \geq x_n$. When the number of measurements is big (over 25–30), it is more convenient to group the results in class intervals. In this case the data is entered into a *primary table*.

When choosing the width of the class interval *b*, the following requirements should be met:

1. It should be the same for all intervals, which facilitates the determination of the statistical estimates.
2. The number of class intervals should fall within $6 \leq k \leq 25$.

3. The approximate number of intervals can be determined according to the following formula:

$$k = 1 + 3.2 \lg n, \quad (I.3.1)$$

where n is the number of measurements.

4. The width of the interval is $b = \frac{x_{max} - x_{min}}{k}$.

It is recommended to round off the value of the interval width in order to facilitate calculations, if they are done manually, or the perception thereof as a whole. For example, if the calculations result is b = 1.89 or b = 2.14, it is preferable to operate with a class interval width of b = 2. In this case the limits of the class intervals will be 0–2, 2–4, 4–6, and so on. For the first case (b = 1.89) they would be 0–1.89, 1.89–3.78, 3.78–5.67, and so on, and for the second one (b = 2.14) 0–2.14, 2.14–4.28, and so on.

The number of results in a given interval is called *absolute frequency*, h_j, and the ratio of the absolute frequency to the volume of the sample n is called *relative frequency*:

$$f_j = \frac{h_j}{n} 100. \quad (I.3.2)$$

When grouping the results together in class intervals, the data is processed under the assumption that all values determined in the respective interval equal the mean of the interval. If we are interested not in the share participation of each interval but in the share of the results above or under a certain value, then what is used is the *cumulative frequency*, Σf_j. It represents the sum of the relative frequencies from the initial to a given class interval.

Example

A measurement should be conducted in order to determine the strength of a certain yarn by performing 120 tests. The expected values fall within the (190–300) cN interval.

The approximate number of class intervals is determined according to (I.3.1):

$$k = 1 + 3.2 \lg 120 = 7.65.$$

The resulting value is rounded off to 8.

The width of the class interval is determined:

$$b = \frac{300 - 190}{8} = 13.75 \text{ cN}.$$

The class width that is selected for operational convenience is close to the calculated one. In order to allow for more precise calculations, the interval is reduced to 10 cN. Table I.3.1 shows the processing order for the grouped data.

The number of the interval is entered in column 1, the interval limits in column 2, the mean of the interval in column 3, the number of occurrences in each interval in column 4, the absolute frequency h_j (the sum of the occurrences) in column 5, the relative frequency f_j in column 6, and the cumulative frequency Σf_j [as determined according to formula (I.3.2)] in column 7.

Table I.3.1. Calculation of the data for drawing of the histogram, polygon and cumulative curve

Class interval	Interval limits	Mean of the interval	Stroke	Absolute frequency h_j	Relative frequency f_j	Cumulative frequency Σf_j
(1)	(2)	(3)	(4)	(5)	(6)	(7)
1	190–200	195	II	2	1.67	1.67
2	200–210	205	II	2	1.67	3.33
3	210–220	215	~~IIII~~ II	7	5.83	9.17
4	220–230	225	~~IIII~~ ~~IIII~~ II	12	10.00	19.17
5	230–240	235	~~IIII~~ ~~IIII~~ ~~IIII~~ III	18	15.00	34.17
6	240–250	245	~~IIII~~ ~~IIII~~ ~~IIII~~ ~~IIII~~ ~~IIII~~ III	28	23.33	57.50
7	250–260	255	~~IIII~~ ~~IIII~~ ~~IIII~~ ~~IIII~~ III	23	19.17	76.67
8	260–270	265	~~IIII~~ ~~IIII~~ IIII	14	11.67	88.33
9	270–280	275	~~IIII~~ ~~IIII~~	10	8.33	96.67
10	280–290	285	III	3	2.50	99.17
11	290–300	295	I	1	0.83	100.00
				120	100	

The test results are graphically represented through:

- A histogram of the frequency distribution (Figure I.3.1)—a diagram consisting of rectangles with a width equal to the width of the class interval and a height equal to the absolute or relative frequencies for the respective interval.
- A distribution polygon (Figure I.3.1)—a curve that represents the relationship between the mean values of the intervals and their absolute or relative frequencies. It is an estimate of the density distribution law.

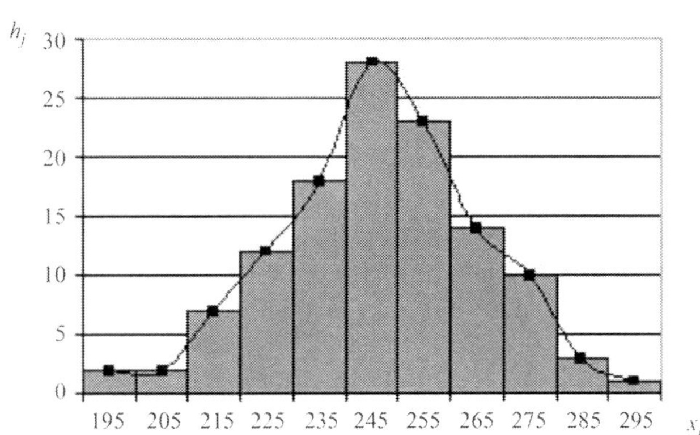

Figure I.3.1. Histogram and polygon.

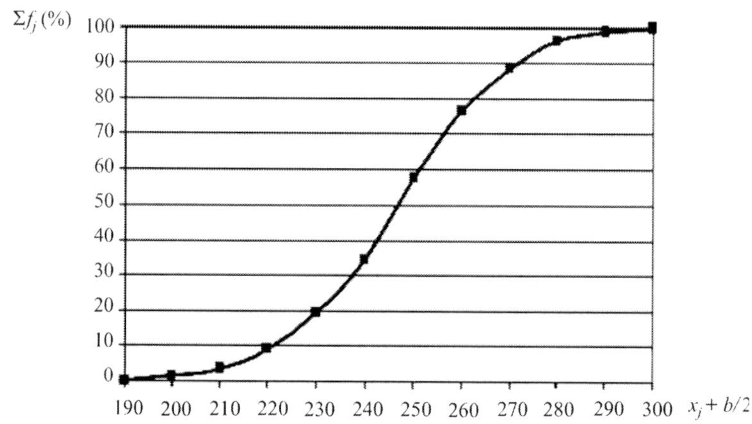

Figure I.3.2. Cumulative curve.

- A cumulative curve (Figure I.3.2)—a curve that represents the probability of the occurrence of a value under the upper limit of the respective class interval. It is an estimate of the distribution function.

With the increase in the size of the sample and narrowing of the interval width, the frequency distribution polygon gets closer and closer to the law on random variable distribution and the cumulative characteristic closer to the distribution function.

POINT ESTIMATES

1. *Arithmetic mean* $\overline{x}$—an estimation of the mathematical expectation.

   It is calculated for:

   - single values

$$\overline{x} = \frac{\sum_{i=1}^{n} x_i}{n}, \tag{I.3.3}$$

where x_i is ith consecutive value,
n is the number of tests.

   - grouped values

$$\overline{x} = \frac{\sum_{j=1}^{k} h_j x_j}{n} = \sum_{j=1}^{k} f_j x_j, \tag{I.3.4}$$

where x_j is the mean of the jth interval,
h_j is the absolute frequency of the jth interval,
f_j is the relative frequency of the jth interval,
k is the number of class intervals.

2. *Mode/Modal value* $\hat{x}$—the estimate of the mode/modal value of the population.

   It is determined as follows:

   - for single values—it corresponds to the value that has the highest frequency;
   - for grouped values

$$\hat{x} = x_m + b\frac{h_m - h_{m-1}}{2h_m - h_{m-1} - h_{m+1}}, \qquad \text{(I.3.5)}$$

where x_m is the upper limit of the modal interval (the interval with the highest frequency),
b is the width of the class intervals,
h_m is the absolute frequency of the modal interval,
h_{m-1} is the absolute frequency of the premodal interval,
h_{m+1} is the absolute frequency of the postmodal interval.

3. *Median* $\tilde{x}$—the estimate of the mode/modal value of the population. The median divides the arranged range in two—the number of values before and after being equal. Geometrically, the median divides the area under the polygon in two equal parts. It is determined as follows:

   - for single values

$$\tilde{x} = x_{\frac{n+1}{2}} \text{ if } n \text{ is an odd number,} \qquad \text{(I.3.6)}$$

$$\tilde{x} = \frac{1}{2}\left(x_{\frac{n}{2}} + x_{\frac{n+1}{2}}\right) \text{ if } n \text{ is an even number} \qquad \text{(I.3.7)}$$

   - for grouped values

$$\tilde{x} = x_{\text{med}} + b\frac{0.5n - \sum_{j=1}^{j_{\text{med}-1}} h_j}{h_{\text{med}}}, \qquad \text{(I.3.8)}$$

where x_{med} is the lower limit of the median interval (the first interval in which the absolute sum frequency is higher than $0.5n$),
$j_{\text{med}-1}$ is the consecutive number of the premedian interval,
h_{med} is the absolute frequency of the median interval.

INTERVAL ESTIMATES

1. *The range R* is an estimate which is very easy to calculate but it serves only as a dispersion indication:

$$R = x_{\max} - x_{\min}. \qquad \text{(I.3.9)}$$

2. *Mean linear deviation d* is an estimate that does not reflect the deviating results to a sufficient degree. It is determined in the following manner:

 - for single values

$$d = \frac{\sum_{i=1}^{n} |x_i - \bar{x}|}{n}, \tag{I.3.10}$$

 where $\bar{x}$ is the arithmetic mean;

 - for grouped values

$$d = \frac{\sum_{j=1}^{k} h_j |x_j - \bar{x}|}{n}. \tag{I.3.11}$$

3. The *variance* S^2 is the estimate of the population variance $D\{x\}$ and is determined as follows:

 - for single values

$$S^2 = \frac{1}{n-1} \sum_{i=1}^{n} (x_i - \bar{x})^2; \tag{I.3.12}$$

 - for grouped values

$$S^2 = \frac{1}{n-1} \sum_{j=1}^{k} h_j (x_j - \bar{x})^2. \tag{I.3.13}$$

 It is a displacement estimate which, due to its adjusted value, is calculated by dividing the sum of the squares of the deviation by $(n - 1)$. The *standard deviation S* is determined on the basis of this estimate:

$$S = \sqrt{S^2}. \tag{I.3.14}$$

 Just like the range and the mean linear deviation, the standard deviation has the dimension of the tested variable but eliminates their defects. It is used both as a variance estimate and as an estimate for the error that occurs due to the incomprehensive nature of the research. This deviation can be used to determine the sustainability of the manufacturing process, and so on. If the size of the sample is small and there is little dispersion in the results (e.g., when testing fabrics), the standard deviation can be calculated with the help of the range as follows:

$$S = \frac{R}{a_n}, \tag{I.3.15}$$

 where a_n is a coefficient which can be taken from Table I.3.2 in accordance with the size of the sample n.

Table I.3.2. Values of the coefficient a_n in accordance with the number of trials n

n	2	3	4	5	6	7	8	9	10	15	20	25
a_n	1.128	1.693	2.059	2.326	2.534	2.704	2.847	2.970	3.078	3.472	3.735	3.931

The described interval estimates have some disadvantages, namely:

- They bear the dimension of the investigated variable and therefore it is not possible to compare the degrees of dispersion of two different indicators, for example, whether the wool fibers show greater dispersion in length or in diameter.
- They do not depend on the mean value. When the mean value is low even the smallest deviations are of importance. For example, if there are two yarns with mean strengths of 100 cN and 350 cN, respectively, and the same standard deviation $S = 20$ cN. At 100 cN, the standard deviation is 20% of the mean value and at 350 cN it is 5.7%.

These disadvantages are eliminated with the so-called *irregularity estimates*. In their nature there are interval estimates (for dispersion) but are calculated as a percentage of the mean value.

4. *Linear irregularity coefficient H*

$$H = \frac{d}{\bar{x}}100. \qquad \text{(I.3.16)}$$

It is used for the comparison of two empirical distributions in view of their dispersion around the mean. Since it is calculated from the mean linear deviation, it has certain drawbacks.

Example
There are two different yarns, each having been measured 100 times. For the first yarn there are 99 results of 100 cN and one of 1 cN. For the second yarn there are 50 measurements of 101 cN and 50 measurements of 97 cN. In both cases the average strength is 99 cN and the linear irregularity is 2%. In the first case, due to the presence of a weak section the yarn will break, and in the second it will not.

5. *Coefficient of variation v*

$$v = \frac{S}{\bar{x}}100. \qquad \text{(I.3.17)}$$

The coefficient of variation is preferable when analyzing the irregularity as it is calculated on the basis of the standard deviation in which the deviations from the mean value are at square degree. Thus the weak deviations die out and the strong ones intensify. In the example above, the coefficient of variation in the first case is 10%, and in the second is 2%.

When the sample is big in size and the law of distribution is normal, the coefficient of variation can be defined by the linear irregularity coefficient using the following formula:

$$v = \sqrt{\frac{\pi}{2}} \cdot H \approx 1.253H. \tag{I.3.18}$$

When the size of the sample is small and the law of distribution is normal, the relationship is as follows:

$$v = A \cdot H, \tag{I.3.19}$$

where A is a coefficient, which can be taken from Table I.3.3 in accordance with the size of the sample n.

Table I.3.3. Values of the coefficient A in accordance with the number of trials n

n	4	6	8	10	15	20	50
A	1.447	1.373	1.340	1.321	1.291	1.286	1.266

CONFIDENCE INTERVALS

The determination of confidence intervals is necessary due to the substitution of the unknown real values with estimates calculated on the basis of a sample. For the establishing of confidence intervals such a distribution is used, the law of which does not depend on unknown values but only on the number of tests n and on the law of distribution of the random variable.

CONFIDENCE INTERVALS OF THE ESTIMATES IN CASES OF NORMAL DISTRIBUTION

1. Mean value

   If the random variable X is normally distributed, the confidence interval of its mathematical expectation is determined as follows:

$$\overline{x} \pm q, \tag{I.3.20}$$

where q is the absolute confidence error. It is determined with the help of the following formula:

$$q = t \cdot \frac{S}{\sqrt{n}}, \tag{I.3.21}$$

where t is the value of Student's t-distribution, which is determined according to the selected level of significance a and the degrees of freedom f (Figure I.3.3):

$$t(a, f = n-1). \tag{I.3.22}$$

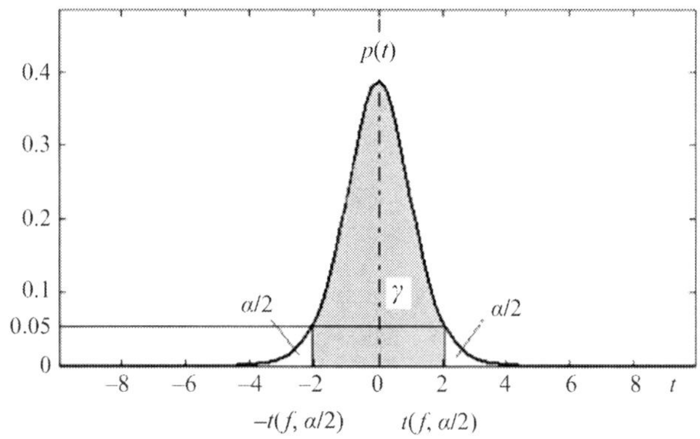

Figure I.3.3. Student's t-distribution.

Student's t-distribution is tabulated and therefore the notion of the table value of Student's t-distribution (criteria) is used very often. The values of the distributions are given in Appendix 1.

As can be seen from formula (I.3.21), the absolute confidence error has the dimension of the investigated value and depends on the standard deviation of the sample, on its size, and on the value of Student's t-distribution. It increases with an increase in the dispersion and an increase in the confidence probability $\gamma = 1 - a$ and decreases with an increase in the number of tests.

The confidence probability is selected in accordance with the desired number of cases within the resulting confidence interval. In the textile practice, people most often use confidence probability of 0.95, which corresponds to a 0.05 level of significance. In certain cases, higher values of the confidence probability can be used.

The confidence interval can also be arrived at through the relative confidence error. It is defined by relating the absolute confidence error to the mean value, respectively, if instead of the standard deviation the variation coefficient is used:

$$p = t \cdot \frac{S}{\bar{x}\sqrt{n}} 100 = t \frac{v}{\sqrt{n}}. \tag{I.3.23}$$

With the help of these formulae one can establish the approximate number of tests $n_{\min}$ for acquiring the set maximum confidence error percentage $p_{\max}$:

$$n_{\min} = \left(\frac{t \cdot v}{p_{\max}} \right)^2. \tag{I.3.24}$$

2. Variance
   When defining the confidence interval of the variance, the Chi-squared (χ^2) distribution is used (Figure I.3.4).

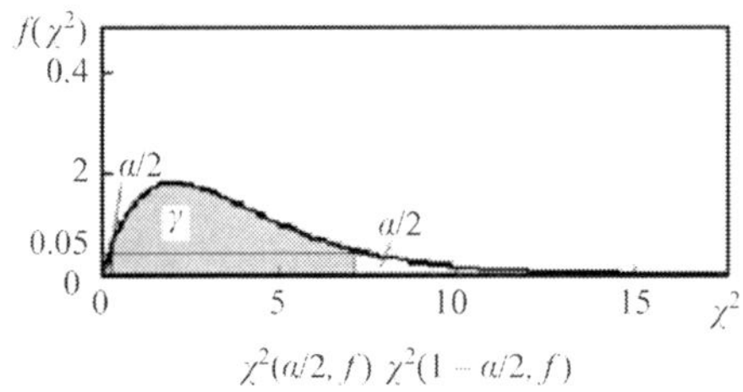

Figure I.3.4. Chi-squared distribution.

Chi-squared distribution is asymmetrical (it tends to symmetrical with the increase in the number of tests) and therefore the confidence level of variance is determined by the following inequality:

$$\frac{(n-1)S^2}{\chi^2(a/2,n-1)} \leq \sigma^2 \leq \frac{(n-1)S^2}{\chi^2(1-a/2,n-1)}, \tag{I.3.25}$$

where the degrees of freedom are $f = n - 1$.

CONFIDENCE INTERVAL OF THE λ PARAMETER FOR POISSON DISTRIBUTION

When the value of the parameter λ ($\lambda > 20$) is big, the distribution tends to normal and therefore the confidence interval of the mathematical expectation can be calculated in a manner similar to that of the normal distribution:

$$\bar{x} \pm t \cdot \sigma. \tag{I.3.26}$$

As in Poisson distribution $\sigma = \sqrt{m_x}$, then:

$$\bar{x} \pm t \cdot \sqrt{\bar{x}}. \tag{I.3.27}$$

In the classical case ($\lambda < 20$), the limits of the confidence interval of the mathematical expectation are:

$$\bar{x} + \frac{t^2}{2} - \sqrt{\frac{t^2}{4} + \bar{x}} \leq m_x \leq \bar{x} + \frac{t^2}{2} + \sqrt{\frac{t^2}{4} + \bar{x}}, \tag{I.3.28}$$

where the value of Student's t-distribution is taken into consideration in the following case $t\,(a, \infty)$. Since $\frac{t^2}{4} << \bar{x}$, this member is very often neglected and the following equation is used in order to determine the confidence interval:

$$\overline{x}+\frac{t^2}{2}-\sqrt{\overline{x}} \leq m_x \leq \overline{x}+\frac{t^2}{2}+\sqrt{\overline{x}}, \qquad (I.3.29)$$

which shows that the interval is symmetrical, however not in reference to the mean value but in reference to $\overline{x}+\frac{t^2}{2}$.

CONFIDENCE INTERVAL OF THE PARAMETER p FOR BINOMIAL DISTRIBUTION

Very often the parameter p cannot be mathematically calculated like dice-throwing. So one should use tests to determine it. If n is the total number of tests, and m is the cases when the desired event occurred, then the ratio m/n in a sufficiently big number of tests is the probability estimate p. In this case the confidence interval is

$$p=\frac{m}{n}\pm t\sqrt{\frac{m}{n^2}\left(1-\frac{m}{n}\right)}. \qquad (I.3.30)$$

If the probability p can be calculated, the confidence interval is

$$\frac{m}{n}-t\sqrt{\frac{p(p-1)}{n}} \leq p \leq \frac{m}{n}+t\sqrt{\frac{p(p-1)}{n}}. \qquad (I.3.31)$$

Example

When calculating the mass of knit sweaters, the following values have been acquired (in g): 200, 210, 215, 210, 210, 195, 200, 190, 210, 205. Determine the statistical estimates.

The arranged range is: 190, 195, 200, 200, 205, 210, 210, 210, 210, 215.

The mean arithmetic value is:

$$\overline{x}=\frac{190+195+200+200+205+210+210+210+210+215}{10}=204.5 \text{ g}.$$

The modal value is $\hat{x}=20$ g as it has the highest rate of occurrence (four times), and the median is determined as the mean arithmetic value of the fifth and the sixth value $\tilde{x}=\frac{205+210}{2}=207.5$ g.

The range is $R=x_{\max}-x_{\min}=215-195=20$ g.

The mean linear deviation is $d=\frac{|190-204.5|+|195-204.5|+2\cdot|200-204.5|+|205-204.5|}{10}$

$$+\frac{4\cdot|210-204.5|+|215-204.5|}{10}=6.6 \text{ g}.$$

(Continued)

(*Continued*)

The variance is $S^2 = \frac{1}{9}\Big[(190-204.5)^2+(195-204.5)^2+2\cdot(200-204.5)^2+(205-204.5)^2$

$+\,4\cdot(210-204.5)^2+(215-204.5)^2\Big]=63.61\ \text{g}^2$

and the standard deviation $S=\sqrt{S^2}=7.98$ g.

The coefficient of variation is $v=\frac{204.5}{7.98}100=3.9\%$.

In order to determine the confidence errors from Appendix 1 at a significance level of $\alpha = 0.05$ and degree of freedom $f = 10 - 1 = 9$, the tabular values of Student's t-distribution are taken into account $t = 2.26$.

The absolute confidence error is $q=\frac{2.26\times 7.98}{\sqrt{10}}=5.7\,\text{g}$.

The percentage confidence error is $p=\frac{2.26\times 3.9}{\sqrt{10}}=2.8\%$.

The confidence interval for the sweaters' mass can be recorded, both, with the absolute and the percentage confidence error:

$$m = (204.5 \pm 5)\ \text{g}$$

$$m = 204.5\ \text{g} \pm 2.8\%$$

It shows that when conducting new measurements, 95% ($\gamma = 1 - 0.05$) of the measured sweaters will have a mass in the interval (195.5–209.5) g.

According to (I.3.21), the confidence interval of the variance can also be determined:

$$\frac{9\times 63.61}{16.92}\le\sigma^2\le\frac{9\times 63.61}{3.33},$$

that is, with a confidence probability $\gamma = 0.95$, the variance lies within [33.84; 171.92] g^2.

Example

In 50 tests, 10 samples proved to be without defects. Find the approximate confidence interval of pulling out a sample without defect.

The probability of pulling out a sample without a defect can be determined approximately as the ratio between the number of samples without defects m and the total number of samples n:

$$\frac{m}{n}=\frac{10}{50}=0.2.$$

The confidence interval is calculated according to (I.3.30), the value of Student's t-distribution being determined from Appendix 1 at a level of significance $\alpha = 0.05$ and degree of freedom $f = \infty$. At $t = 1.96$, the confidence interval is $p = [0.0891, 0.3109]$.

I.4. Statistical Process Control and Control Charts

STATISTICAL PROCESS CONTROL

The statistical process control (SPC) is applied in established technological processes in order to observe, analyze, and regulate them. The use of SPC allows for timely warning in case of a deviation and, if necessary, to correct the parameters of the process. The application of SPC is a prerequisite for the timely detection of the reasons behind quality deterioration. It helps in decreasing the defects, maintaining the quality at a selected level, and improving the technological process (Figure I.4.1).

The control is realized through regular inspections carried out according to a certain schedule that uses the methods of mathematical statistics (Figure I.4.2).

The correct running of the technological processes is assessed with the *coefficients of process capability*.

The coefficient of process stability, C_p, represents the relationship between the tolerance of the controlled parameter T and the area of dispersion V:

$$C_p = \frac{T}{V} = \frac{T_u - T_l}{6 \cdot S}, \tag{I.4.1}$$

where S is the standard deviation of the parameter,

T_u and T_l are the upper and lower limits of the parameter tolerance, respectively.

When the dispersion area is lower than the tolerance, ($C_p > 1$), the technological process is stable and can be controlled by statistical methods. The control of this process can help in avoiding defective products. When $C_p < 1$, the process is unstable and the occurrence of defects is inevitable. In order to make this process manageable it is necessary to make changes in the technology, to repair or replace the machines. A value over 1.33 is considered a good coefficient of process stability as it guarantees the stability of the process for a longer period. For a short period the stability is determined mainly by the machine, while for a longer period other factors may play a role, such as wearing out, changes in the parameters of the environment or in the parameters of the material.

Apart from being stable, a process has to be set in the right manner. The setting of the technological processes is characterized by the so-called *coefficient of adjustment*, C_{pk}, which

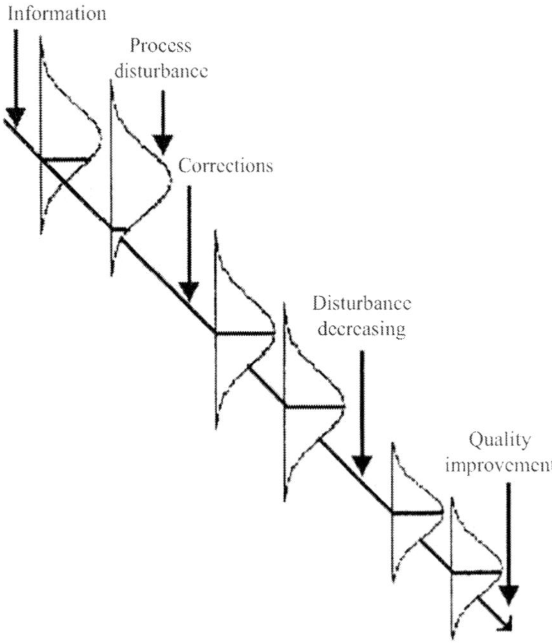

Figure I.4.1. Steps for quality improvement.

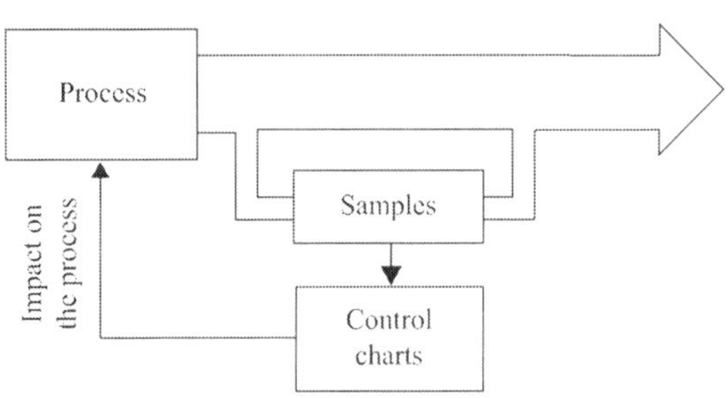

Figure I.4.2. Principle of SPS.

defines the displacement of the center of the grouped results toward the middle of the tolerance field. Since the displacement can occur in both directions two coefficients are determined, C_{pk1} and C_{pk2}, the coefficient of adjustment being the smaller of the two:

$$C_{pk1} = \frac{T_l - \overline{X}}{3 \cdot S}, \tag{I.4.2}$$

$$C_{pk2} = \frac{\overline{X} - T_u}{3 \cdot S}, \tag{I.4.3}$$

$$C_{pk} = \left|\min\left(C_{pk1}, C_{pk2}\right)\right|. \quad (I.4.4)$$

The coefficient of adjustment C_{pk} should have a value of more than 1. Otherwise, defects can occur in the production, and their percentage depends on the type of distribution. For a normally distributed random variable, the values of the defective production dependent on the coefficient of adjustment are given in Table I.4.1.

Table I.4.1. Percentage of defective production for a normally distributed random variable

C_{pk}	0	0.1	0.2	0.3	0.4	0.5	0.6	0.7	0.8	0.9	1
Defective production (in %)	50	38	27	18	12	7	4	2	1	0.4	0

Figure I.4.3 represents the typical cases of distribution of the controlled variable under the normal law of distribution [(a) set but unstable process, (b) set and stable process, (c) stable process with a borderline setting, (d) stable but unset process].

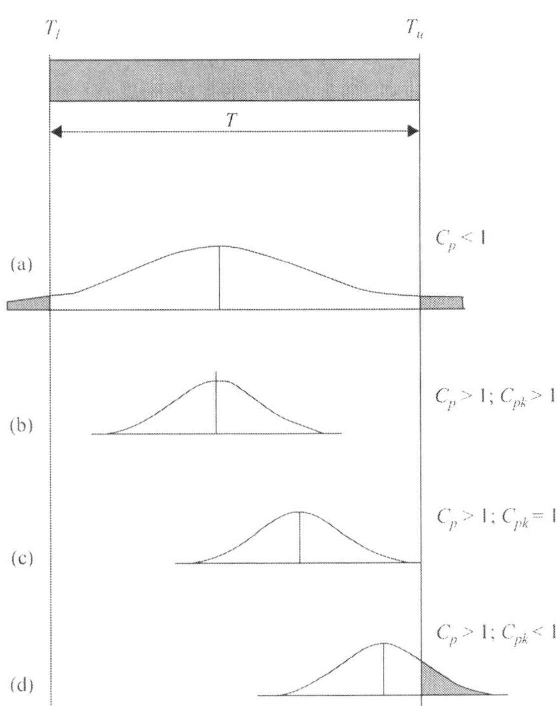

Figure I.4.3. Some typical cases of distribution of the controlled variable.

CONTROL CHARTS

The operational control is implemented with the help of control charts—a piece of cardboard on the x-axis of which are recorded the discrete moments of time in which the control has been performed or the number of the sample (observation), and on the y-axis is the controlled indicator.

The statistical hypothesis can be verified graphically by the control chart. The confidence interval is built for a certain estimate θ under the condition that the zero hypothesis is being met H_0: $\theta = \theta_0$, where θ_0 is the value desired by us or required by a certain standard or a document. The zero hypothesis is not rejected if the estimate of the specific parameter falls within the limits of the confidence interval, and it is rejected in the opposite case. Most often, the numerical characteristics of the distribution of random variables (mathematical expectation, median, variance, range, and so on) are used for estimates.

The creation of a control chart is based on the following prerequisites: the controlled property is a random value and, therefore, the individual values vary in a certain interval:

$$\mu \pm t \cdot \sigma, \qquad (I.4.5)$$

where μ is the mathematical expectation of the random value,
σ is the standard deviation,
t is Student's t-distribution value at a selected confidence probability.

Most often, the confidence probability used is $\gamma = 0.99$ which corresponds to $t = 2.576$ or $\gamma = 0.9973$ which corresponds to $t = 3$. The second case is used more often since the value of t is a whole number and the percentage of case coverage is almost 100.

The control chart consists of three lines—one central that corresponds to the nominal or mean value and two parallel ones—the upper and lower control limits, which set the width of the confidence interval at the selected confidence probability (Figure I.4.4). Sometimes on the control chart are also positioned two warning lines at a distance of $\pm 2S$.

Keeping a control chart means to record the value of the investigated property on the chart in selected moments. If there are values that fall outside the interval, defined on the basis of a preliminary sample, it means that the technological process is disturbed and an intervention is necessary in order to regulate it. If only one value falls outside the interval, this does not imply a correction in the process since the interval set does not cover 100% of the cases (when $\gamma = 0.99$ it is permissible for 1 of a 100 values to fall outside the set limits). If there is a value outside the

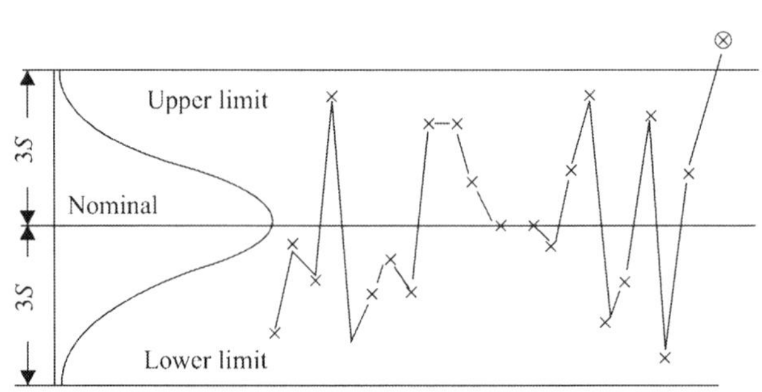

Figure I.4.4. Graphical presentation of the control chart limits.

control limits defined by the standardization requirements (through the limits of the tolerance field), this is already a signal for the appearance of defects. All products manufactured after the last regular sample should be subjected to overall control.

DESIGN OF STATISTICAL CONTROL CHARTS

When designing a control chart it is necessary to comply with the following requirements:

- the controlled variable should be random; and
- the controlled technological process should be stable ($C_p > 1$).

Data from the preliminary measurement is used in chart design with the objective of finding the mean and the standard deviation. For the standard deviation to be close enough to that of the population, 100 to 200 primary data are necessary. The primary data should be taken from different packages or in different times which guarantees that the sample is representative.

Example
When determining the limits of the control chart for yarn strength, it is preferable to make 20 measurements from five spools than 100 measurements from one spool.

The next task is to establish whether the resulting values are from one and the same population. This could be checked with the help of Student's criteria by comparing the mean values of the different packages. More often, the approach of excluding the strongly deviating results is used. The method includes calculating the mean and the standard deviation of the primary data and establishing the confidence interval. All results that fall outside the limits of the confidence interval are eliminated (Figure I.4.5). New interval limits are calculated with the remaining values until all values fall within the newly calculated confidence interval.

Control charts can also be designed on the basis of standardization of requirements, including nominal value, tolerance, upper and lower tolerance limits, and acceptable quality level (AQL). AQL is the maximum medium-level of defectiveness which is considered to be

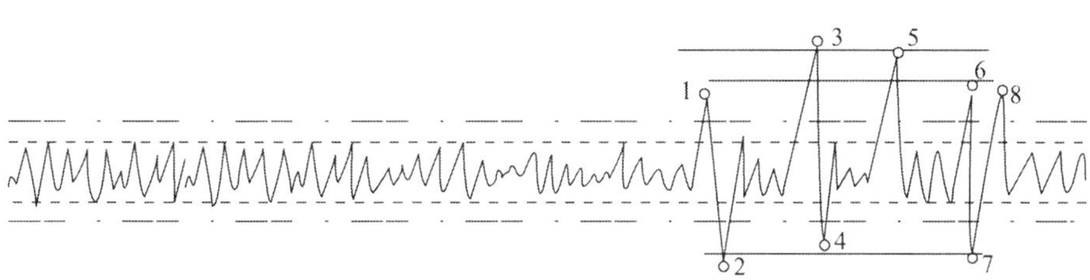

Figure I.4.5. Algorithm for calculation of the control limits.

acceptable. It is a matter of agreement, but as a reference it is selected within the following limits in dependence to the type of product and its responsibility:

- AQL = 0.04–0.4%—for products with a high level of responsibility such as those used in the healthcare and defense sector.
- AQL = 0.65–1.5%—for products with high level of responsibility used in technical facilities.
- AQL = 2.5–6.5%—mass technical tasks (such as those in the textile production).
- AQL = 6.5–15%—products with low level of responsibility.

TYPES OF CONTROL CHARTS

- for measurable (quantitative) indicators; and
- for countable (qualitative) indicators.

If possible, the control charts for quantitative indicators should always be preferred as they provide better accuracy.

Control Charts for Quantitative Indicators

For this type of control charts, the time is recorded along the x-axis and the qualitatively measurable indicators about the quality of the production are along the y-axis. When setting the control limits it is necessary to know the type of distribution of the random variable. The most widely spread types of control charts are those for normally distributed random variables. In case the distribution of the random variable differs from the normal one, other suitable distributions or free of distribution criteria might be used for verification of hypotheses.

In order to find out the distribution type, usually a preliminary sample is made which consists of a large number of elements (no less than 150–200 observations). These observations are used for the creation of a histogram and then visually it is estimated whether the distribution can be considered to be a normal one. If necessary, a statistical verification of the distribution type can be performed.

If it is necessary to control two or more indicators, those indicators could be checked for independence with the help of the procedures for examination of hypothesis for coefficients of correlation. If two indicators are interrelated then only one of them should be used as it contains information about the other one. Thus the statistical control is greatly facilitated.

On the basis of the preliminary sample the capability of the process is defined, such as its adjustment and stability. The charts for qualitative indicators with an established normal law of distribution have the biggest practical application. According to the Bulgarian State Standard BDS 11319:1990 *Statistical quality regulation. Control charts* the following combined charts are used:

- $\bar{x} / S$ chart for mean arithmetic value and standard deviation;
- $\bar{x} / R$ chart for mean arithmetic value and range;
- $\tilde{x} / R$ chart for median and range;
- Control chart for the practical limits;
- Control chart for individual values and absolute values of the subsequent differences $x_i / |x_i - x_{i+1}|$.

Initially the chart that characterizes the dispersion of the process is analyzed. After the standard deviation or the range is stabilized, the chart that characterizes the setting of the process is analyzed.

$\overline{x}$ / S CONTROL CHART

The chart is effective if the size of the sample is bigger than 8. It is convenient for usage when there are tools for the quick calculation of $\overline{x}$ and S. What is necessary to calculate the charts are $k = 20$–30 samples with a total number of single tests $n = 1000$–200.

The mean arithmetic value $\overline{x}_i$ and the standard deviation S_i are established for all samples:

$$\overline{x}_i = \frac{1}{n}\sum_{j=1}^{n} x_{ij}, \tag{I.4.6}$$

$$S_i = \sqrt{\frac{1}{n-1}\sum_{j=1}^{n}(x_{ij} - \overline{x}_i)^2}. \tag{I.4.7}$$

The resulting values are recorded on the control charts and the individual points are connected with a broken line that shows the trends in the change of certain characteristics.

The mean values of $\overline{x}_i$ and S_i are determined by

$$\overline{\overline{x}} = \frac{1}{k}\sum_{i=1}^{k} \overline{x}_i, \tag{I.4.8}$$

$$\overline{S} = \frac{1}{k}\sum_{i=1}^{k} S_i. \tag{I.4.9}$$

The resulting values are recorded on the charts (Figure I.4.6).

The control limits K_l and K_u are determined according to the formulae given in Table I.4.2, and the values of the coefficients A_3, A_6, B_3, B_4, B_5, and B_6 are given in Table I.4.3.

When the control chart is calculated according to the standardization requirements, the limits are calculated in relation to the tolerance field T and the standard deviation S_0, as defined in big samples or according to the selected acceptable quality level (Table I.4.4).

The calculated control limits are recorded in the chart (Figure I.4.6).

On the variance chart only the upper limit is represented. The value of the lower limit has no practical significance since the movement of the results in its direction is a positive sign that the process is stabilized.

$\overline{x}$ / R CONTROL CHART

The chart is similar to the $\overline{x}$ / S chart but instead of the standard deviation the range is used to estimate the dispersion. The range for each sample is calculated:

$$R_i = x_{i,\max} - x_{i,\min}, \tag{I.4.10}$$

Table I.4.2. Calculation of the control limits for $\bar{x}$ / S control chart

Type of chart	Control limits: According to test data	Control limits: According to standardization requirements	Comment
$\bar{x}$	$K_u = \bar{\bar{x}} + A_3 \cdot \bar{S}$ $K_l = \bar{\bar{x}} - A_3 \cdot \bar{S}$	$K_u = T_u - A_6 (T/2)$ $K_l = T_l + A_6 (T/2)$	In combination with S-chart
S	$K_u = B_4 \cdot \bar{S}$ $K_l = B_3 \cdot \bar{S}$	$K_u = B_6 \cdot S_0$ $K_l = B_5 \cdot S_0$	When $n \leq 5$ $K_l = 0$

Table I.4.3. Auxiliary coefficients for calculation of the control limits

Coefficients	n = 2	3	4	5	6	7	8	9	10
A_2	1.880	1.023	0.729	0.577	0.483	0.419	0.373	0.337	0.308
A_3	2.659	1.954	1.628	1.427	1.287	1.182	1.099	1.032	0.975
A_5	2.800	1.360	0.950	0.740	0.620	0.540	0.480	0.430	0.390
A_6	0.293	0.423	0.500	0.553	0.592	0.622	0.646	0.667	0.684
B_3	0	0	0	0	0.030	0.118	0.185	0.239	0.284
B_4	3.267	2.568	2.266	2.089	1.970	1.882	1.815	1.761	1.716
B_5	0	0	0	0	0.029	0.113	0.179	0.232	0.276
B_6	2.606	2.276	2.088	1.964	1.874	1.806	1.751	1.707	1.669
D_1	0	0	0	0	0	0.204	0.388	0.547	0.687
D_2	3.686	4.358	4.698	4.918	5.078	5.204	5.306	5.393	5.469
D_3	0	0	0	0	0	0.076	0.136	0.184	0.223
D_4	3.267	2.574	2.282	2.114	2.004	1.924	1.864	1.816	1.777
d_2	1.128	1.693	2.059	2.326	2.534	2.704	2.847	2.970	3.078

Table I.4.4. Standard deviation in accordance to the selected AQL

AQL (in %)	0.04	0.065	0.10	0.15	0.25	0.40	0.65
S_0	$T/7$	$T/6.8$	$T/6.6$	$T/6.4$	$T/6$	$T/5.78$	$T/5.44$
AQL (in %)	1.0	1.5	2.5	4.0	6.5	10.0	15.0
S_0	$T/5.16$	$T/4.86$	$T/4.46$	$T/4.12$	$T/3.7$	$T/3.3$	$T/2.88$

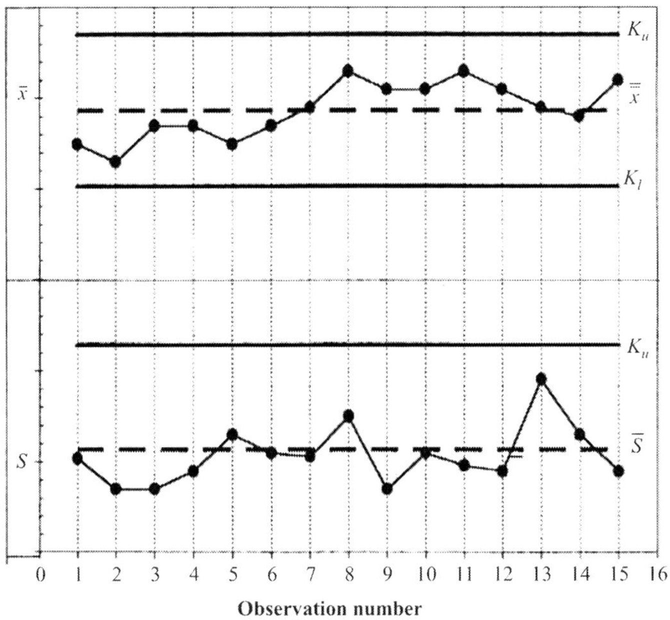

Figure I.4.6. $\bar{x}$ / S control chart.

and the individual values are used for the calculation of the mean range of all k number of samples:

$$\bar{R} = \frac{1}{k}\sum_{i=1}^{k} R_i. \tag{I.4.11}$$

The range is a less effective estimate in comparison with the standard deviation but is easier in terms of calculation. The formulae for the calculation of the control limits are given in Table I.4.5.

Table I.4.5. Calculation of the control limits for $\bar{x}$ / S control chart

Type of card	Control limits: According to test data	Control limits: According to standardization requirements	Comment
$\bar{x}$	$K_u = \bar{\bar{x}} + A_2 \cdot \bar{R}$ $K_l = \bar{\bar{x}} - A_2 \cdot \bar{R}$	$K_u = T_u - A_6(T/2)$ $K_l = T_l + A_6(T/2)$	In combination with R-chart
R	$K_u = D_4 \cdot \bar{R}$ $K_l = D_3 \cdot \bar{R}$	$K_u = D_2 \cdot S_0$ $K_l = D_1 \cdot S_0$	When $n \le 6$ $K_l = 0$

Very often, with time the mean value changes under the influence of a certain factor, for example, the wearing out of the card-clothing wire leads to an increase in the number of neps in the card sliver. In these cases an $\bar{x}$ / R chart with a "moving average" which allows the retrieval of information, both about the accidental disturbances and the effect of the dominant factor, is used.

The chart of mean arithmetic value is developed on the basis of the sample data taken during the period between two changes of the card-clothing wire. After determining and recording the mean arithmetic values of the individual samples on the chart, the approximating line M is drawn (Figure I.4.7).

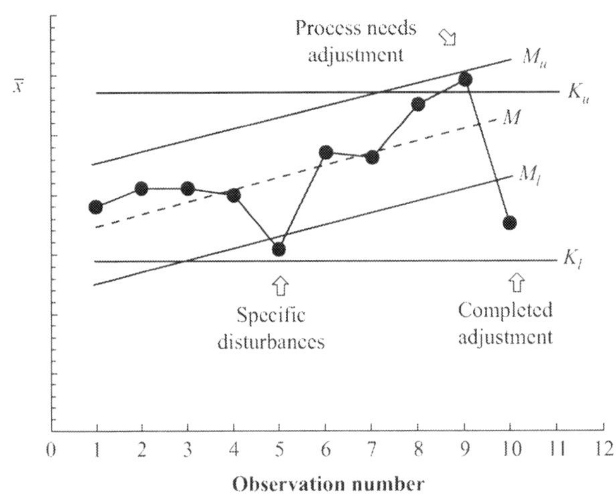

Figure I.4.7. $\bar{x}$ / R chart with a "moving average".

Parallel to the line M are drawn the limits M_l and M_u, located at a distance of $A_2 \cdot \bar{R}$ from the line. If a value falls outside these limits, this is a signal for the occurrence of specific disturbances in the process. The horizontal lines K_l and K_u are also recorded on the chart and they are determined in accordance with the following:

$$K_{u,l} = \bar{\bar{x}} \pm \left(\frac{\Delta \bar{x}}{2} + A_2 \bar{R} \right), \qquad \text{(I.4.12)}$$

where $\Delta\bar{x}$ is the change in the mean values for the studied period. If the mean value falls outside these limits, this is a signal that it is necessary to adjust the settings or replace the tool.

The range chart is the same as in the preceding case.

$\tilde{x}$ / R CONTROL CHART

The chart is similar to the $\bar{x}$ / R chart but the median is used for the estimation of the center of the grouped results and its determination is much easier.

When working with these types of charts it is recommended that the number of measurements is an odd number. If the measurements are an even number, the median is determined as the mean arithmetic values of the two central values. The median value $\tilde{x}_i$ is determined for every sample. The mean median is determined on the basis of the median values:

$$\bar{\tilde{x}} = \frac{1}{k}\sum_{i=1}^{k} \tilde{x}_i. \qquad (I.4.13)$$

The control limits are determined according to the formulae given in Table I.4.6.

Table I.4.6. Calculation of the control limits for $\tilde{x} / R$ control chart

	Control limits		
Type of chart	**According to test data**	**According to standardization requirements**	**Comment**
$\tilde{x}$	$K_u = \bar{\tilde{x}} + A_5 \cdot \bar{R}$ $K_l = \bar{\tilde{x}} - A_5 \cdot \bar{R}$	$K_u = T_u - 0.8A_6(T/2)$ $K_l = T_l + 0.8A_6(T/2)$	
R	$K_u = D_4 \cdot \bar{R}$ $K_l = D_3 \cdot \bar{R}$	$K_u = D_2 \cdot S_0$ $K_l = D_1 \cdot S_0$	When $n \le 6$ $K_l = 0$

The mean value of the median and the control limits is recorded on the chart. The range chart is calculated in the same way as for the $\bar{x} / R$ chart.

The standard deviation of the median is

$$S_{\tilde{x}} = \frac{S}{\sqrt{n}}\sqrt{\frac{\pi}{2}}, \qquad (I.4.14)$$

that is, with $\sqrt{\frac{\pi}{2}} = 1.25$ being bigger than the mean arithmetic value. Since the variance of the median is 25% higher, the control limits are wider ($A_5 > A_2$), which results in less effective control.

CONTROL CHART FOR PRACTICAL LIMITS

This chart is designed for statistical regulation of parameters with normal distribution and a two-sided tolerance field with limits T_u and T_l. Typically in this chart, the control limits change in a dynamic way because they are calculated on the basis of the practical limits of dispersion. They are determined for each sample in accordance with its size n and probability γ for the inclusion of a certain percentage P of the values of the measured parameter in the control limits. The control limits $T_{u,i}$ and $T_{l,i}$ are calculated according to the following formulae:

$$K_{u,i} = \overline{x}_i + K \cdot S_i \qquad \text{(I.4.15)}$$

$$K_{l,i} = \overline{x}_i - K \cdot S_i, \qquad \text{(I.4.16)}$$

where K is a coefficient dependent on n, γ, and P.

It can be taken from tables or determined with the help of Student's t-distribution $t(\gamma, n-1)$, which reflect the probability γ for inclusion of a given percentage of the values of the measured parameter in certain limits:

$$K = \frac{t}{\sqrt{n}} \qquad \text{(I.4.17)}$$

P and γ are most often assumed to equal 0.90, 0.95, and 0.99.

For each sample, on the control chart are recorded the points $\overline{x}_i$, $K_{u,i}$, and $K_{l,i}$, the limits T_u and T_l, and the median T_M of the tolerance field (Figure I.4.8).

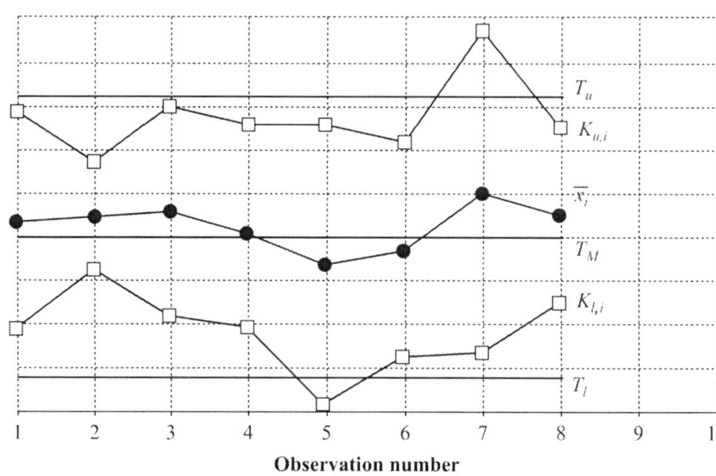

Figure I.4.8. Control chart for practical limits.

The location of points $K_{u,i}$ and $K_{l,i}$ serves as a basis for the dispersion and the location of points $\overline{x}_i$ as the basis for the adjustment of the process.

The controlled process is considered to be statistically manageable and capable when, during a given period of time, corresponding to more than five consecutive samples, the points $K_{u,i}$ and $K_{l,i}$ are located in the tolerance field between T_u and T_l.

CONTROL CHART FOR THE INDIVIDUAL AND ABSOLUTE VALUES OF THE SEQUENTIAL DIFFERENCES $x_i / | x_i - x_{i+1} |$

This type of control chart is applicable in the production of small batches. The sequential differences are determined by $|x_1 - x_2|$, $|x_2 - x_3|$, and so on. The mean of the individual values and the range are determined according to the following formulae:

$$\bar{\bar{x}} = \frac{1}{k}\sum_{i=1}^{k} x_i \qquad (I.4.18)$$

and

$$\bar{R} = \frac{1}{k}\sum_{i=1}^{k} |x_i - x_{i+1}|, \qquad (I.4.19)$$

where k is the number of individual values. The control limits are determined according to Table I.4.7, and the coefficients according to Table I.4.3 when $n = 2$.

Table I.4.7. Calculation of the control limits for control chart for the individual and absolute values of the sequential differences

Type of chart	Control limits: According to test data	Control limits: According to standardization requirements	Comment
x_i	$K_u = \bar{\bar{x}} + 3\bar{R} / d_2$ $K_l = \bar{\bar{x}} - 3\bar{R} / d_2$	$K_u = T_u - 3S_0$ $K_l = T_l + 3S_0$	$n = 2$
$\|x_i - x_{i+1}\|$	$K_u = D_4 \cdot \bar{R}$ $K_l = D_3 \cdot \bar{R}$	$K_u = D_2 \cdot S_0$ $K_l = D_1 \cdot S_0$	$n = 2$

For control of the manufacturing process for the production of small batches, the following types of control charts can be used:

- chart for small batches of similar products; and
- chart for small batches with different distribution characteristics.

CONTROL CHARTS FOR BATCHES OF SIMILAR PRODUCTS

Very often a certain machine is used for the consecutive production of small batches of similar products. For example, a spinning machine could be used for the production of yarns with linear density of 20 and 25 tex or a knitting machine could make products of different sizes. In this case, it is more appropriate to use relative deviations instead of absolute values, that is, the control charts for mean value are designed according to a zero mean line.

The procedures for the design of these charts are identical with those for classical charts. The difference is that the limits of the different batches of the products are recorded on the control charts.

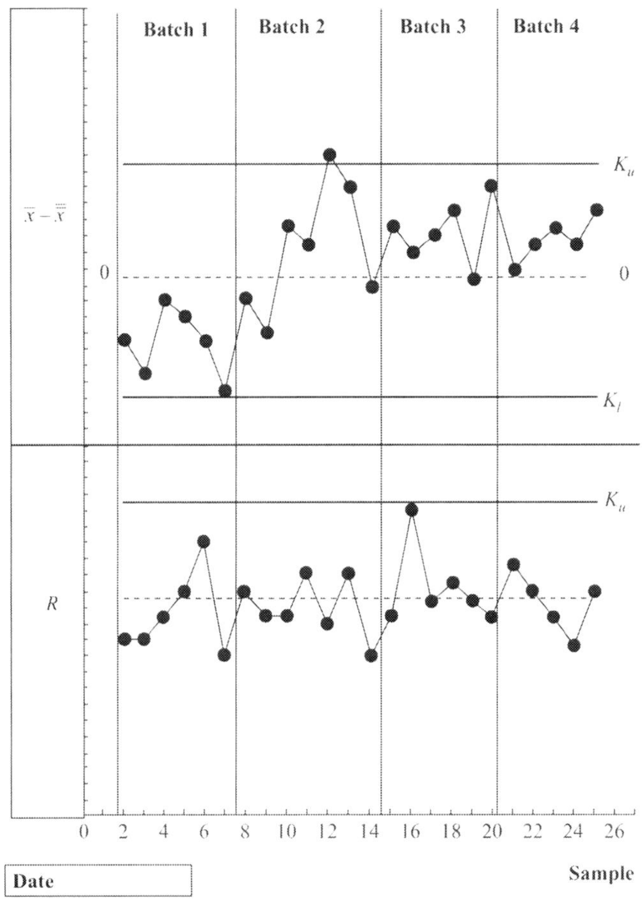

Figure I.4.9. Control charts for batches of similar products.

The sequence in the design of charts is the following:

1. The mean value $\bar{\bar{x}}$ of the controlled parameter is determined for every sample. The mean line on the control chart is marked as zero (Figure I.4.9).
2. The mean value is determined for every sample with size n and the differences are recorded on the chart as $\Delta\bar{x} = \bar{x} - \bar{\bar{x}}$. The values for the sample range are recorded on the range chart.
3. The control limits are determined. For the mean values chart: $K_{u,l} = \pm A_2\bar{R}$. For the range chart: $K_u = D_4\bar{R}$.
4. The limits for the individual batches are recorded distinctly on the control limits (Figure I.4.9).

CONTROL CHARTS FOR SMALL BATCHES WITH DIFFERENT DISTRIBUTION CHARACTERISTICS

Very often, the batches differ not only in terms of the mean value but also in terms of the degree of variance. This is typical for such cases when the difference in the mean values is big or when one and the same product is made from different materials. For example, when there is a big difference in the linear density of the yarn, the deviations in the absolute values are also big. At the same time, yarns with the same linear density but made from cotton or polyester fibers have a different variance, that is, the control limits of the range should be different. This requires a transformation of the regular $\overline{x} / R$ charts. For regular charts, the values of $\overline{x}$ should fall within the following limits

$$\overline{\overline{x}} - A_2\overline{R} < \overline{x} < \overline{\overline{x}} + A_2\overline{R}. \quad (I.4.20)$$

If $\overline{\overline{x}}$ is subtracted and the equation is divided by 3, the result is the following equivalent equation:

$$-A_2 < \frac{\overline{x} - \overline{\overline{x}}}{\overline{R}} < A_2. \quad (I.4.21)$$

Thus, the permanent control limits $\pm A_2$ can be recorded on the control chart for the mean values. Such a transformation can be made for each batch.

In a similar manner, the inequality for the range:

$$D_3\overline{R} < R < D_4\overline{R} \quad (I.4.22)$$

can be transformed into:

$$D_3 < \frac{R}{\overline{R}} < D_4, \quad (I.4.23)$$

which allows the determination of constant control limits for the range that are defined by the coefficients D_3 and D_4.

The control charts of this type are designed according to the following algorithm:

1. The mean value $\overline{\overline{x}}$ is determined for every batch. It can be determined on the basis of a preliminary investigation; when such information is not available or is insufficient it can be determined on the basis of the nominal value.
2. The mean value of the range $\overline{R}$ is determined. When there is no reliable information about the mean value of the range, it is determined by selecting the desired value of the coefficient of process capability C_p:

$$\overline{R} = \frac{T \cdot d_2}{6C_p}. \quad (I.4.24)$$

3. For each sample the value of $\frac{\bar{x}-\bar{\bar{x}}}{\bar{R}}$ is determined and the resulting values are recorded on the "mean values" chart according to the zero line.
4. The control limits for the "mean values" $\pm A_2$ are determined.
5. For each sample, the ratio $R / \bar{R}$ is determined and recorded on the range chart.
6. The control limits for the range, D_3 and D_4, are determined.
7. The limits between the separate batches are recorded.

In cases of small batches, the classical control charts can also be used but instead of the mean values of the individual samples, the individual values are recorded.

The variance of singular values S is higher than the variance of the mean values $\frac{S}{\sqrt{n}}$ (Figure I.4.10). It is a special case when $n = 1$.

Several other charts are applicable in the textile practice on the basis of the specific features of the production process.

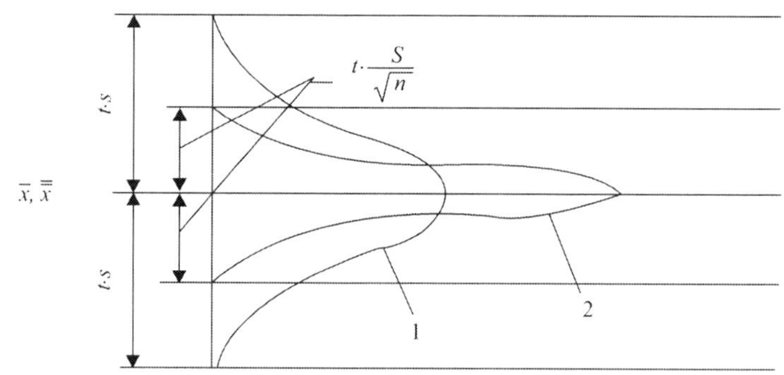

Figure I.4.10. Control charts for small batches with different distribution characteristics.

CONTROL CHARTS FOR GROUP VALUES (*i*-CHARTS)

They are used when several measurements are made for one sample, for example, in the control of the linear density of the yarn. The equality of the mean values and the variances should be made in advance (according to Student's and Fisher criteria). The resulting values are recorded one above the other and it should be observed whether some of the points do not fall outside the limits of the confidence interval (Figure I.4.11).

The confidence interval is determined in the same way as for the chart for individual values. The *i*-chart is a chart for primary data and the chart for the individual values is a special case of this chart when $n = 1$.

CONTROL CHARTS FOR EXTREME VALUES (*k*-CHARTS)

They are a special case of the group value charts. From the resulting n values, the only ones that are recorded are the two extreme values—x_{min} and x_{max}. Since they fall within the control

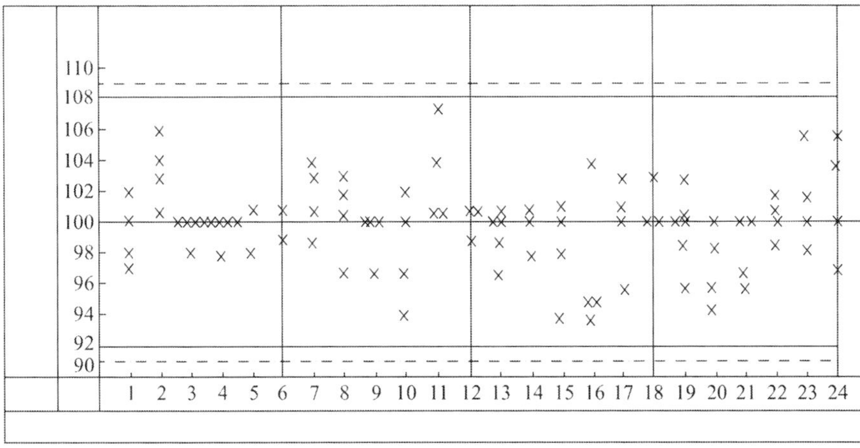

Figure I.4.11. Control charts for group values.

limits, it infers that the other values also fall within them. For better clarity the points are joined together with broken lines (Figure I.4.12). The control limits are calculated in the same way as for the group values.

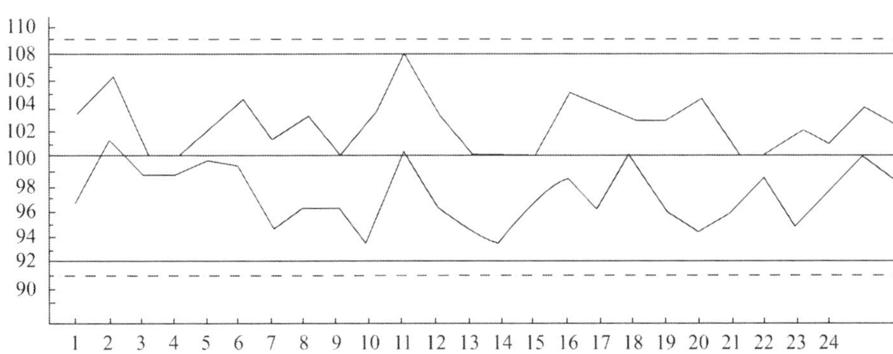

Figure I.4.12. Control charts for extreme values.

Example
Determine the control limits for the main statistical estimates of the linear density of cotton sliver. Five measurements have been made in a period of 20 days and the linear density has been recorded in Nm. For facilitation, the values have been multiplied by 1000, that is. when the measured value has been Nm 0.254, the value recorded in the table is 254. The data from the measurements is represented in Table I.4.8, columns (2)–(6).

(Continued)

Table I.4.8. Data and statistical estimates for the example

Observation day			Values			$\bar{x}$	S	R	$\tilde{x}$
(1)	(2)	(3)	(4)	(5)	(6)	(7)	(8)	(9)	(10)
1	254	252	250	252	256	252.8	2.28	6	252
2	258	258	256	256	252	256	2.45	6	256
3	256	250	252	254	252	252.8	2.28	6	252
4	252	246	252	258	254	252.4	4.34	12	252
5	244	250	246	254	248	248.4	3.85	10	248
6	248	246	258	254	254	252	4.90	12	254
7	252	254	252	250	252	252	1.41	4	252
8	252	256	252	256	250	253.2	2.68	6	252
9	248	256	258	252	252	253.2	3.90	10	252
10	260	258	260	250	252	256	4.69	10	258
11	254	256	250	246	252	251.6	3.85	10	252
12	251	247	240	244	244	245.2	4.09	11	244
13	250	248	250	244	248	248	2.45	6	248
14	258	254	263	254	250	255.8	4.92	13	254
15	254	258	248	252	252	252.8	3.63	10	252
16	250	252	244	250	252	249.6	3.29	8	250
17	252	250	252	246	248	249.6	2.61	6	250
18	250	250	252	252	250	250.8	1.10	2	250
19	254	260	252	270	272	261.6	9.10	20	260
20	250	250	276	276	276	265.6	14.24	26	276
		Mean values				252.97	4.10	9.7	253.2

In columns (7), (8), (9), and (10) are calculated the mean value, the standard deviation, the range, and the median, respectively, of the results for different days.

CALCULATION OF $\bar{x}$ / S CONTROL CHART

The control limits for the mean values are determined according to the formulae given in Table I.4.2:

$$\text{For the lower limit: } K_l = \bar{\bar{x}} - A_3 \cdot \bar{S} = 252.97 - 1.427 \times 4.1 = 247.116;$$

$$\text{For the upper limit: } K_u = \bar{\bar{x}} + A_3 \cdot \bar{S} = 252.97 + 1.427 \times 4.1 = 258.824.$$

It should be checked if there are mean values that fall outside the set limits. Three cases fall outside the set limits: Day number 12, under the set lower limit, and Day numbers 19 and 20, above the set limit (Figure I.4.13).

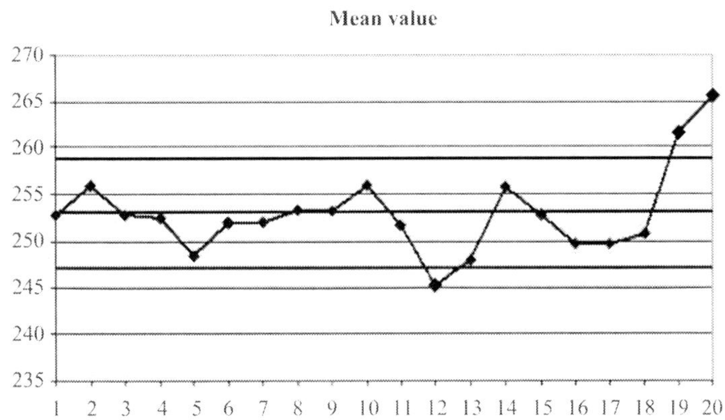

Figure I.4.13. $\overline{x}$-chart during its calculation.

The limits are calculated again and the quoted values are excluded. The new values of the mean of the mean values and the mean standard deviation are $\overline{\overline{x}} = 252.176$ and $\overline{S} = 3.21$ and the new control limits are:

For the lower limit: $K_l = \overline{\overline{x}} - A_3 \cdot \overline{S} = 252.176 - 1.427 \times 3.21 = 247.592;$

For the upper limit: $K_u = \overline{\overline{x}} + A_3 \cdot \overline{S} = 252.176 + 1.427 \times 3.21 = 256.761.$

When the limits are calculated in this manner, there are no cases of values that fall outside them and therefore, it can be inferred that these are the final values of the control limits for the mean value.

In a similar manner the control limits for the standard deviation are calculated by excluding the cases that have already been excluded in the determination of the control limits for the mean value (Day numbers 12, 19, and 20).

For the lower limit: $K_l = B_3 \cdot \overline{S} = 0 \times 3.21 = 0;$

For the upper limit: $K_u = B_4 \cdot \overline{S} = 2.089 \times 3.21 = 6.71.$

These limits for the standard deviation are final since there are no cases that fall outside them. The $\overline{x} / S$ control chart is drawn with the help of the already calculated control limits (Figure I.4.14).

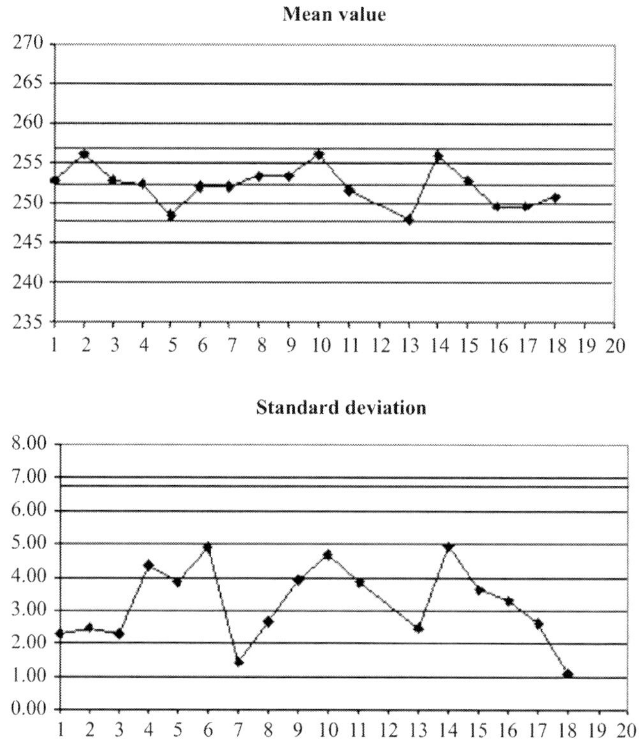

Figure I.4.14. Final $\overline{x}$ / S chart for the example.

CALCULATION OF THE $\overline{x}$ / R CONTROL CHART

The calculation of the control limits for the mean value and the range is done according to Table I.4.5. For the mean value, the control limits are:

$$\text{For the lower limit: } K_l = \overline{\overline{x}} - A_2 \cdot \overline{R} = 252.97 - 0.577 \times 9.7 = 247.37;$$

$$\text{For the upper limit: } K_u = \overline{\overline{x}} + A_2 \cdot \overline{R} = 252.97 + 0.577 \times 9.7 = 258.57.$$

Again, several cases fall outside the calculated limits—Day numbers 12, 19, and 20. The adjusted limits are as follows:

$$\text{For the lower limit: } K_l = \overline{\overline{x}} - A_2 \cdot \overline{R} = 252.176 - 0.577 \times 8.06 = 247.53;$$

$$\text{For the upper limit: } K_u = \overline{\overline{x}} + A_2 \cdot \overline{R} = 252.176 + 0.577 \times 8.06 = 256.83,$$

and after the abovementioned cases are disregarded, $\overline{R} = 8.06$. These are final since there are no cases that fall outside them.

In regard to the range, the control limits are:

For the lower limit: $K_l = D_3 \cdot \overline{R} = 0 \times 8.06 = 0$;

For the upper limit: $K_u = D_4 \cdot \overline{R} = 2.114 \times 8.06 = 17.04$.

These are also final. They serve as a basis for drawing the control charts presented in Figure I.4.15.

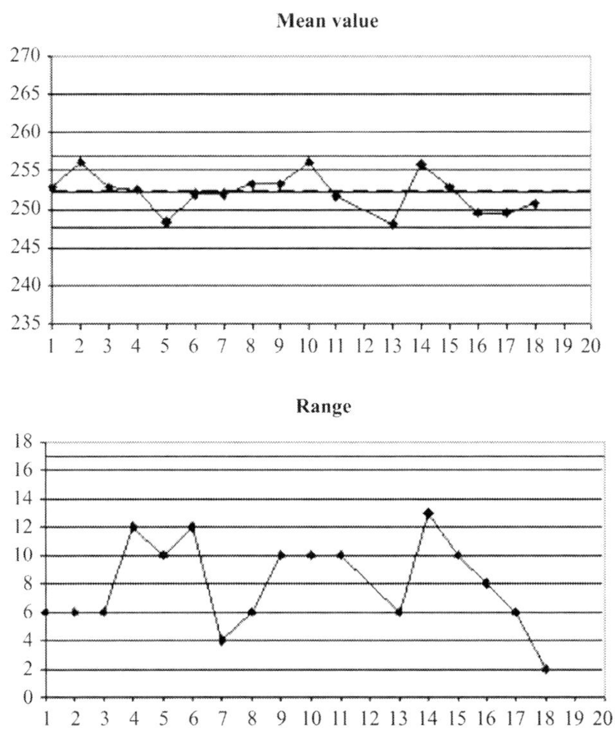

Figure I.4.15. Final $\overline{x}$ / R control chart for the example.

In a similar manner, the other types of control charts can be calculated and drawn.

Control Charts for Qualitative Indicators

In numerous cases, the values of the controlled parameter can only be measured in a qualitative and not in a quantitative manner. For example, presence/absence of a defect, good for sale/ defective. In order to implement current statistical control, the so-called control charts for immeasurable (qualitative) indicators are used in these cases. In comparison with the charts for quantitative indicators, these are simpler and can be used to control several parameters and the costs for control, design, and filling in are significantly lower. Along with these advantages, they have several disadvantages: they are less sensitive to the changes in the technological processes and provide less information.

There are four types of control charts for qualitative indicators:

- control chart for the relative number of defective products (*p*-chart);
- control chart for the number of defective products (*np*-chart);
- control chart for the number of defects (*c*-chart); and
- control chart for the relative number of defects (*u*-chart).

CHART FOR THE RELATIVE NUMBER OF DEFECTIVE PRODUCTS (*p*-CHART)

The chart is based on the calculation of the relative number of defective products in the sample:

$$p_i = \frac{f_i}{n_i}. \tag{I.4.25}$$

where f_i is the number of defective products,
n_i is the number of the products in the sample.

The difference in the size of the samples should not be higher than 25% and the minimum sample size is determined according to the following condition:

$$n_{\min} \geq \frac{4}{\overline{p}} \quad \text{or} \tag{I.4.26}$$

$$n_{\min} \geq \frac{4}{AQL}. \tag{I.4.27}$$

where $\overline{p}$ is the mean arithmetic value of the relative number of defective products and defines the location of the central line on the chart. It is determined as follows:

$$\overline{p} = \frac{1}{m}\sum_{i=1}^{m} p_i, \tag{I.4.28}$$

where m is the number of samples.

The number of samples should not be less than 20, and in order to set the control charts correctly, the size of the samples should be big enough—approximately 100–200 products. The control limits are calculated according to the formulae given in Table I.4.9, the mean size of the sample being:

$$\overline{n} = \frac{n_1 + n_2 + \cdots + n_m}{m}. \tag{I.4.29}$$

When calculating the control limits according to the standardization requirements, the AQL is taken into consideration. The standardization requirements are compatible with the accuracy capabilities of the process, if $AQL \geq \overline{p}$ and the statistical regulation of the technological process is easy to implement. The standardization requirements are difficult to achieve when $AQL < \overline{p}$. In this case it is necessary to improve the processes.

Table I.4.9. Calculation of the control limits for *p*-chart

	Control limits	
Type of chart	**According to test data**	**According to standardization requirements**
p-chart	$K_u = \overline{p} + 3\sqrt{\frac{\overline{p}(1-\overline{p})}{\overline{n}}}$ $K_l = \overline{p} - 3\sqrt{\frac{\overline{p}(1-\overline{p})}{\overline{n}}}$	$K_u = AQL + 3\sqrt{\frac{AQL(1-AQL)}{\overline{n}}}$ $K_l = AQL - 3\sqrt{\frac{AQL(1-AQL)}{\overline{n}}}$

The calculated mean line and the control limits are recorded on the control chart (Figure I.4.16).

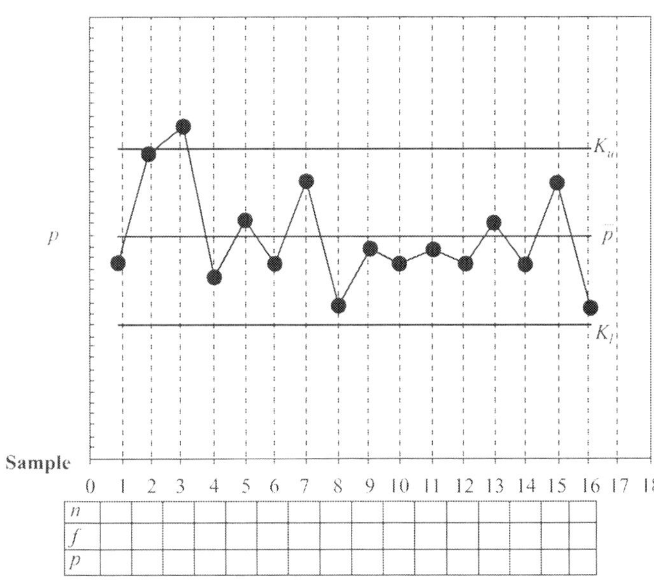

Figure I.4.16. *p*-chart.

If the control is 100% then the control limits are calculated everyday on the basis of the volume of the daily production n_i:

$$K_u = \overline{p} + 3\sqrt{\frac{\overline{p}(1-\overline{p})}{n_i}}, \tag{I.4.30}$$

$$K_l = \bar{p} - 3\sqrt{\frac{\bar{p}(1-\bar{p})}{n_i}}. \qquad (I.4.31)$$

In this case the control limits are not straight but broken lines (Figure I.4.17).

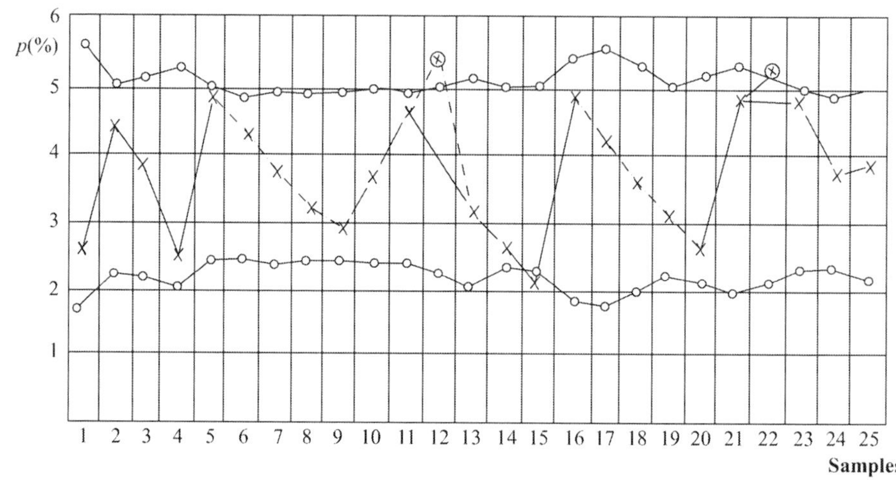

Figure I.4.17. *p*-chart in the case of 100% control.

CHART FOR THE NUMBER OF DEFECTIVE PRODUCTS (*np*-CHART)

The control *np*-charts resemble the *p*-charts, but for them it is necessary that the number of samples, *n*, is the same as the actual number of defective products, f_i, in the sample is recorded on the chart. The control limits are calculated according to the formulae given in Table I.4.10 where the mean defectiveness of the products in the samples is

$$\bar{f} = \frac{f_1 + f_2 + \cdots + f_m}{m}. \qquad (I.4.32)$$

Table I.4.10. Calculation of control limits for *np*-chart

Type of chart	Control limits: According to test data	Control limits: According to standardization requirements
np-chart	$K_u = n\bar{p} + 3\sqrt{n\bar{p}(1-\bar{p})}$	$K_u = nAQL + 3\sqrt{nAQL(1-AQL)}$
	$K_l = n\bar{p} - 3\sqrt{n\bar{p}(1-\bar{p})}$	$K_l = nAQL - 3\sqrt{nAQL(1-AQL)}$

If the resulting value for the lower limit is negative then the lower limit is assumed to be zero.

CONTROL CHART FOR THE NUMBER OF DEFECTS (*c*-CHART)

The *p*- and *np*-charts are used for the control of the percentage (part) of defective products. The *c*- and *u*-charts are used to control the number of defects in a group of products (a batch), in one product, or in a preselected notional unit (the number of defects in a roll of fabric, the number of neps in 1g flor, the number of neps in 1000 m yarn, the number of defects in a batch of 50 suits). The *c*-chart is used when the number of products in the batch is one and the same. The control limits are calculated according to the formulae given in Table I.4.11, where $\overline{c}$ is the mean number of defects.

Table I.4.11 Calculation of control limits for *c*-chart

	Control limits	
Type of chart	**According to test data**	**According to standardization requirements**
c-chart	$K_u = \overline{c} + 3\sqrt{\overline{c}}$ $K_l = \overline{c} - 3\sqrt{\overline{c}}$	$K_u = AQL + 3\sqrt{AQL}$ $K_l = AQL - 3\sqrt{AQL}$

The acceptable level of defectiveness is given in the number of defects, *n,* per unit of production or the number of defects per unit of length, area, or volume. The control limits of the *c*-charts are rounded down to a whole number since the number of defects can only be a whole number.

CONTROL CHART FOR THE RELATIVE NUMBER OF DEFECTS (*u*-CHART)

The *u*-chart is used when the number of products in the different batches is different and they should be reduced to a unit of production:

$$u_i = \frac{c_i}{n_i}. \tag{I.4.33}$$

The control limits are calculated according to the formulae given in Table I.4.12 where $\overline{u}$ is the mean number of defects in a unit of production/notional unit, and *n* is the mean size of the samples. The acceptable level of defectiveness is given in number of defects per unit of production.

Figure I.4.18 gives examples of *c*- and *u*-charts.

Table I.4.12. Calculation of control limits for *u*-chart

Type of chart	Control limits	
	According to test data	**According to standardization requirements**
u-chart	$K_u = \bar{u} + 3\sqrt{\frac{\bar{u}}{n}}$ $K_l = \bar{u} - 3\sqrt{\frac{\bar{u}}{n}}$	$K_u = AQL + 3\sqrt{\frac{AQL}{n}}$ $K_l = AQL - 3\sqrt{\frac{AQL}{n}}$

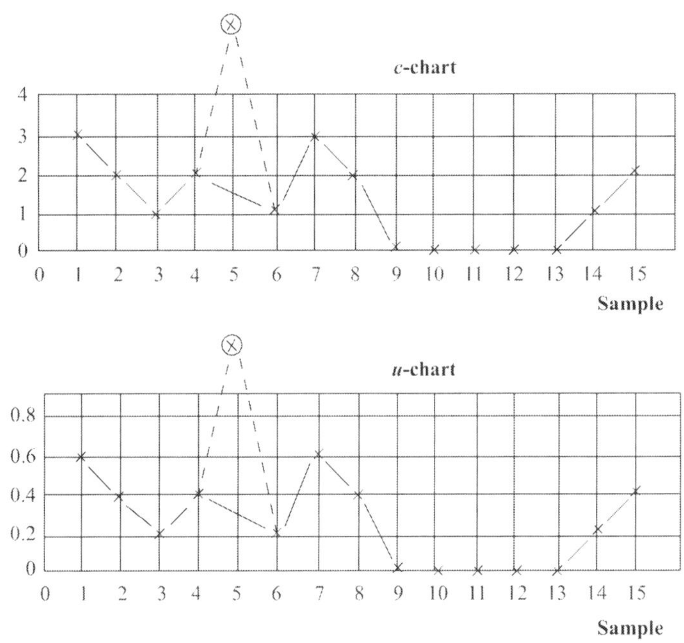

Figure I.4.18. *c*- and *u*-charts.

Example

Design a control chart for controlling the defectiveness in a textile factory. A research has been carried out over a period of 25 days and the data about the products made, n_i, and the defective products, x_i, is given in Table I.4.13.

(Continued)

In column 3 is the calculated percentage of defective products, p_i, for each day. The mean percentage of defective products $\bar{p} = 3.79\%$ and the mean number of products made $\bar{n} = 1573.2$ are determined.

Table I.4.13. Calculation of control limits for *p*-chart

Observation day number	n_i	x_i	p_i	K_u	K_l	K_u	K_l
(1)	**(2)**	**(3)**	**(4)**	**(5)**	**(6)**	**(7)**	**(8)**
1	837	21	2.51	5.77	1.81	5.68	1.76
2	1540	67	4.35	5.25	2.33	5.17	2.27
3	1455	55	3.78	5.29	2.29	5.21	2.23
4	1230	30	2.44	5.42	2.16	5.34	2.10
5	1814	88	4.85	5.14	2.44	5.05	2.39
6	2151	91	4.23	5.03	2.55	4.94	2.50
7	2032	76	3.74	5.06	2.52	4.98	2.46
8	2094	67	3.20	5.04	2.54	4.96	2.48
9	1973	58	2.94	5.08	2.50	5.00	2.44
10	1823	66	3.62	5.13	2.45	5.05	2.39
11	2093	96	4.59	5.04	2.54	4.96	2.48
12	1536	82	5.34	5.25	2.33	—	—
13	1314	41	3.12	5.37	2.21	5.29	2.15
14	1756	46	2.62	5.16	2.42	5.07	2.37
15	1628	37	2.27	5.21	2.37	—	—
16	954	46	4.82	5.64	1.94	5.56	1.88
17	881	37	4.20	5.72	1.86	5.63	1.81
18	1136	41	3.61	5.49	2.09	5.40	2.04
19	1546	47	3.04	5.25	2.33	5.16	2.28
20	1338	36	2.69	5.36	2.22	5.27	2.17
21	1095	54	4.93	5.52	2.06	5.44	2.00
22	1959	103	5.26	5.08	2.50	—	—
23	1850	91	4.92	5.12	2.46	5.04	2.40
24	1813	68	3.75	5.14	2.44	5.05	2.39
25	1482	59	3.98	5.28	2.30	5.19	2.25

The upper and lower control limits for the *p*-chart are calculated according to the formulae given in Table I.4.9 (the *np*-chart is not applicable as the number of products made each day is different).

(*Continued*)

(*Continued*)

Upper control limit:

$$K_u = \overline{p} + 3\sqrt{\frac{\overline{p}(1-\overline{p})}{\overline{n}}} = 0.0379 + 3\sqrt{\frac{0.0379(1-0.0379)}{1573.2}} = 0.0524 = 5.24\%$$

The limits could also be calculated directly in percentage.

Lower control limit:

$$K_l = \overline{p} - 3\sqrt{\frac{\overline{p}(100-\overline{p})}{\overline{n}}} = 3.79 - 3\sqrt{\frac{3.79(100-3.79)}{1573.2}} = 2.35\%$$

Cases 12, 15, and 22 fall outside the control limits. They are taken out and the new control limits are calculated on the basis of the new mean percentage of defective products $\overline{p} = 3.72\%$ and the mean number of products made—$\overline{n} = 1554.9$.

Upper control limit:

$$K_u = \overline{p} + 3\sqrt{\frac{\overline{p}(100-\overline{p})}{\overline{n}}} = 3.72 + 3\sqrt{\frac{3.72(100-3.72)}{1554.9}} = 5.17\%$$

Lower control limit:

$$K_l = \overline{p} - 3\sqrt{\frac{\overline{p}(100-\overline{p})}{\overline{n}}} = 3.72 - 3\sqrt{\frac{3.72(100-3.72)}{1554.9}} = 2.28\%$$

These control limits are final, since there are no cases that fall outside them. The control chart is represented in Figure I.4.19.

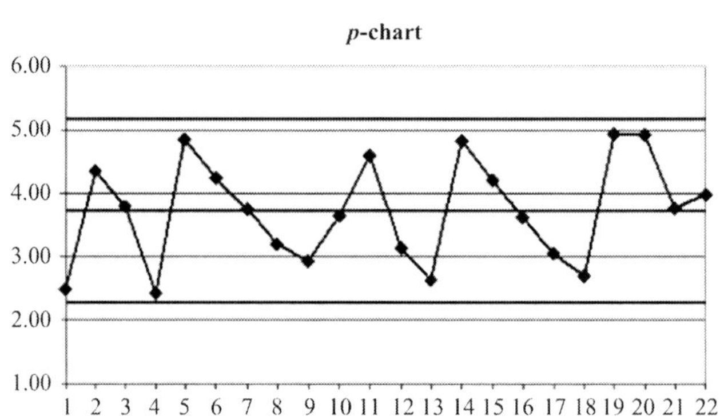

Figure I.4.19. Final version of the *p*-chart for the example.

If the control is not going to be a sample control but a 100% control, the control limits should be calculated according to formulae (I.4.30) and (I.4.31). The calculated values for the upper limits are entered in column 5 of Table I.4.13 and for the lower limits in column 6. As expected, cases 12, 15, and 22 fall outside the control limits. The values calculated after they have been taken out for the upper and lower limits are given, respectively, in columns 7 and 8. There is no case that falls outside the resulting values which allows for the drawing of the control chart, represented in Figure I.4.20.

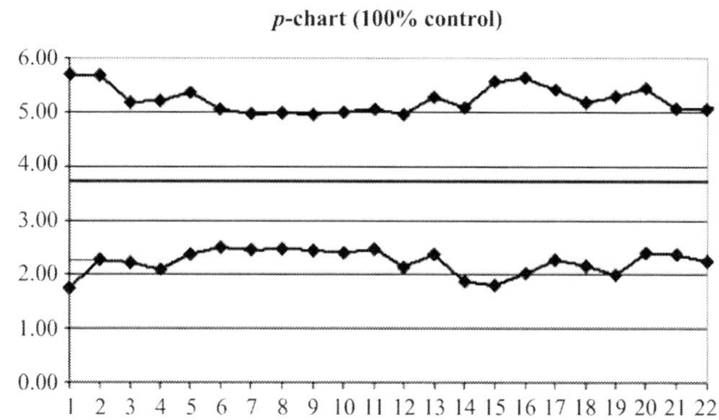

Figure I.4.20. *p*-chart for 100% control for the example.

Example

On the basis of 15 samples, each consisting of 5 products, design the control charts for the number of defects and their relative number per unit of production.

The number of established defects for every sample is given in column (2) in Table I.4.14.

The number of defects per unit of production $u_i = \frac{c_i}{n}$ $(n = 5)$ is calculated in column (3).

The formulae from Table I.4.11 are used to determine the control limits for the number of defects after the mean number of defects have already been established:

$$\bar{c} = \frac{\sum_{i=1}^{15} c_i}{m} = \frac{23}{15} = 1.53 \text{ defects per sample.}$$

Upper control limit:

$$K_u = \bar{c} + 3\sqrt{\bar{c}} = 1.53 + 3\sqrt{1.53} = 5.25$$

It is assumed that $K_u = 5$.

(*Continued*)

(Continued)

Table I.4.14. Number of established defects for the samples

Sample number	c_i	u_i
(1)	**(2)**	**(3)**
1	3	0.6
2	2	0.4
3	1	0.2
4	2	0.4
5	6	1.2
6	1	0.2
7	3	0.6
8	2	0.4
9	0	0
10	0	0
11	0	0
12	0	0
13	0	0
14	1	0.2
15	2	0.4

Lower control limit:

$$K_l = \overline{c} - 3\sqrt{\overline{c}} = 1.53 - 3\sqrt{1.53} = -2.18.$$

It is assumed that $K_l = 0$.

The results of sample number 5 fall outside the control limits and therefore it is taken out and they are calculated anew. The mean number of defects for the other 14 samples is $\overline{c} = 1.21$.

The new control limits are:

Upper control limit:

$$K_u = \overline{c} + 3\sqrt{\overline{c}} = 1.21 + 3\sqrt{1.21} = 4.52.$$

It is assumed that $K_u = 4$.

(Continued)

Lower control limit:

$$K_l = \bar{c} - 3\sqrt{\bar{c}} = 1.21 - 3\sqrt{1.21} = -2.09.$$

It is assumed that $K_l = 0$.

These limits are final. The controlled limits for the *u*-chart are calculated in a similar way. The formulae from Table I.4.12 are used where:

$$\bar{u} = \frac{\sum_{i=1}^{15} c_i}{\sum_{i=1}^{15} n_i} = \frac{23}{15 \times 5} = 0.307 \text{ defects per product.}$$

Upper control limit:

$$K_u = \bar{u} + 3\sqrt{\frac{\bar{u}}{n}} = 0.307 + 3\sqrt{\frac{0.307}{5}} = 1.05$$

Lower control limit:

$$K_l = \bar{u} - 3\sqrt{\frac{\bar{u}}{n}} = 0.307 - 3\sqrt{\frac{0.307}{5}} = -0.80.$$ It is assumed that $K_l = 0$.

After sample number 5 is dropped out, the new mean value is $\bar{u} = 0.243$ defects per product and the new control limits are:

Upper control limit:

$$K_u = \bar{u} + 3\sqrt{\frac{\bar{u}}{n}} = 0.243 + 3\sqrt{\frac{0.243}{5}} = 0.904.$$

Lower control limit:

$$K_l = \bar{u} - 3\sqrt{\frac{\bar{u}}{n}} = 0.243 - 3\sqrt{\frac{0.243}{5}} = -0.75.$$ It is assumed that $K_l = 0$.

These control limits are final.

I.5. Correlation Analysis

The preceding sections reviewed mass events (populations, distributions, and respectively, statistical estimates) of only one indicator. In practice, however, it is very often necessary to study mass events under several indicators, for example, under x and y. In this case a system of two random variables is present. The combined study of these indicators is of interest especially when there is a certain cause-and-effect relationship between them, that is, when the first indicator affects the second one and changes its characteristics in a certain way.

In the area of exact natural science, each value of a certain property (variable) x corresponds to a specific value of another property (variable) y. In the textile practice, however, most often the dependency between the two properties is not exactly a functional, but a statistical one. This means that the total statistical population of y corresponds to a specific value of x and, vice versa, a specific value of y corresponds to the total statistical population of x (Figure I.1.9).

The correlation theory studies the correlation dependencies and, more specifically, it is used to define the power of dependency between the causing event and the resulting event, and respectively, to what extent this dependency approaches the functional one. There are different methods for calculating the qualitative characteristics for the power of the correlation dependency (linear correlation, correlation relationship, and so on). The method is selected in accordance with the following factors:

- How are the cause-and-effect–bound indicators of the studies events expressed—in a descriptive or qualitative way;
- How is the data is presented—is it ungrouped or grouped;
- What is the nature of the cause-and-effect relationship that has to be established—is itlinear or nonlinear dependency;
- What is the number of reasons for which the researcher needs to establish the effect on the consequence—using simple correlation and multiple (double, triple, and so on, respectively specific) correlation.

Typical examples for the application of the correlation theory in the textile practice are the establishment of relationship between the length and the fineness of the fibers, the fineness and the specific mechanical indicators (e.g., breaking strength) between the properties of the fibers and those of the yarns, respectively the fabrics, between the moisture content and the breakage in spinning and weaving, between the strength and the elasticity, between the friction resistance and wearing out resistance (in test wearing), between the warp tension and the breakage in weaving, between the speed of the spindles, looms, and so on, and the breakage, between the component composition and the breakage, respectively the physical–mechanical properties of the yarns and fabrics, and so on.

COEFFICIENT OF LINEAR CORRELATION

When there is expected linear dependency and there are at least three couples of values for x and y, the *coefficient for linear correlation*, r_{xy}, is calculated:

$$r_{xy} = \frac{\sum_{i=1}^{n}\left[(x_i - \overline{x})(y_i - \overline{y})\right]}{\sqrt{\sum_{i=1}^{n}(x_i - \overline{x})^2 \cdot \sum_{i=1}^{n}(y_i - \overline{y})^2}}, \tag{I.5.1}$$

where $\overline{x}$ is the mean arithmetic value of the measurements for the variable x,
$\overline{y}$ is the mean arithmetic value of the results from the measurements of the variable y,
n is the number of tests.

The coefficient of linear correlation determines the strength and the direction of the impact between the random variables and it can have values within the [−1; 1] interval. The significance of its value is clarified in Section I.1, in parameters and numerical characteristics of the random variable. The significance of the coefficient of linear correlation is checked with the help of Student's criterion, the accountable value being determined according to the following formula:

$$t_R = \frac{r_{xy}\sqrt{n-2}}{\sqrt{1-r_{xy}^2}}. \tag{I.5.2}$$

The resulting value is checked against the table value, determined under the condition: $t_T\left(\frac{a}{2}, f = n-2\right)$. The coefficient of linear correlation is of significance if $t_R > t_T$. Otherwise, even if the absolute value is high, it is insignificant.

COEFFICIENT OF DETERMINATION, ANTICORRELATION, AND INDETERMINATE COEFFICIENTS

For characterization of the correlations dependencies in mathematical statistics, other estimates can also be used such as coefficient of determination, anticorrelation coefficient, and indeterminate coefficient.

- The *coefficient of determination* shows what percentage of changes in the random variable y is due to changes in the variable x:

$$R_0 = 100 r_{xy}^2. \tag{I.5.3}$$

- The *coefficient of anticorrelation* characterizes the absence of a link between variable y and variable x:

$$K_0 = \sqrt{1 - r_{xy}^2}. \quad (I.5.4)$$

- The *indeterminate coefficient* shows what percentage of the changes in y is not due to changes in x:

$$S_0 = 100 \cdot K^2 = 100 - R. \quad (I.5.5)$$

Example

Ten tests have been conducted in order to determine the strength x (N) and the elasticity y (%) up to yarn breakage. Find a proof if there is a statistically proven relationship between those two indicators. The data from the measurements is given in Table I.5.1 (columns 1–3).

Table I.5.1. Data and calculation of the linear correlation coefficient

Test number	x_i	y_i	$(x_i - \bar{x})$	$(y_i - \bar{y})$	$(x_i - \bar{x})(y_i - \bar{y})$	$(x_i - \bar{x})^2$	$(y_i - \bar{y})^2$
(1)	**(2)**	**(3)**	**(4)**	**(5)**	**(6)**	**(7)**	**(8)**
1	7.29	7.83	−0.716	−1.234	0.884	0.513	1.523
2	8.20	9.31	0.194	0.246	0.048	0.038	0.061
3	8.75	10.00	0.744	0.936	0.696	0.554	0.876
4	7.73	8.28	−0.276	−0.784	0.216	0.076	0.615
5	8.46	10.00	0.454	0.936	0.425	0.206	0.876
6	7.05	7.40	−0.956	−1.664	1.591	0.914	2.769
7	7.80	9.18	−0.206	0.116	−0.024	0.042	0.013
8	8.37	9.55	0.364	0.486	0.177	0.132	0.236
9	8.42	9.53	0.414	0.466	0.193	0.171	0.217
10	7.99	9.56	−0.016	0.496	−0.008	0.000	0.246
				Σ =	4.198	2.647	7.432

The mean value of the strength is $\bar{x} = 8.006$ N, the mean value of the elasticity is $\bar{y} = 9.064\%$. The value of the *linear correlation coefficient* is determined according to formula (I.5.1), the auxiliary amounts are calculated in columns 4–8:

$$r_{xy} = \frac{4.198}{\sqrt{2.947 \times 7.432}} = 0.95.$$

(Continued)

(*Continued*)

The significance of the linear correlation coefficient is checked with the help of the Student's criterion, the accountable value being determined by following formula (I.5.2):

$$t_R = \frac{0.95\sqrt{10-2}}{\sqrt{1-0.95^2}} = 8.60.$$

The resulting value is checked against the table value, determined under the following conditions:

$$t_T\left(\frac{a}{2} = 0.025, f = 10-2 = 8\right) = 2.75.$$

The coefficient of linear correlation is of significance since $t_R > t_T$. This means that between the strength and the elasticity up to the point of breakage, there is a strongly expressed (the value of the linear correlation coefficient is close to 1) and positive (the increase in the strength leads to increase in the elasticity up to the point of breakage of the yarn) statistical dependence.

The *coefficient of determination* is determined according to formula (I.5.3):

$$R_0 = 100 \cdot r_{xy}^2 = 100 \times 0.95^2 = 90.25.$$

The *coefficient of anticorrelation* is determined according to formula (I.5.4):

$$K_0 = \sqrt{1-r_{xy}^2} = \sqrt{1-0.95^2} = 0.31.$$

The *indeterminate coefficient* is determined according to formula (I.5.5):

$$S_0 = 100 \cdot K_0^2 = 100 - R_0 = 100 \times 0.31^2 = 100 - 90.25 = 9.75.$$

CORRELATION IN CASE OF ALTERNATIVE INDICATORS—THE FOUR-FIELD METHOD

Often the cause and effect cannot be presented in a numerical way but only in a descriptive one, with the help of alternatives. For example, the manufactured production could be standard or nonstandard, the staff qualified or nonqualified. In these cases the degree of causal dependency can be determined with the help of the four-field method. The following formula is used:

$$K = \frac{a \cdot d - b \cdot c}{\sqrt{(a+b)\cdot(c+d)\cdot(a+c)\cdot(b+d)}}, \qquad \text{(I.5.6)}$$

where *a, b, c*, and *d* are the number of the cases for the separate combinations of the cause-and-effect options.

In order to facilitate the calculation of the coefficient K, the so-called quadruple table (Table I.5.2) is used, in which are entered the two possible alternatives of the cause, x_1 and x_2, and of the effect, y_1 and y_2. Index 1 is given for the desired (positive) option, for example, standard production, qualified staff, and index 2 for the undesired (negative) option, for example, defective production, unqualified staff.

Table I.5.2. Quadruple table for calculation of the degree of the causal dependency

X \ Y	y_1	y_2	**Sum**
x_1	a	b	$(a + b)$
x_2	c	d	$(c + d)$
Sum	$(a + c)$	$(b + d)$	$(a + b + c + d)$

With a is indicated the number of cases when the combination x_1y_1 is present and with b the number of cases when the combination x_1y_2 is present, c is the number of cases for the combination x_2y_1 and d is the number of cases for the combination x_2y_2.

It is accepted that the degree of dependence is weak when $K \leq 0.3$, moderate when $0.3 \leq K \leq 0.5$, significant when $0.5 \leq K \leq 0.7$, big when $0.7 \leq K \leq 0.9$, and extremely big when $K \geq 0.9$. The accuracy of the established dependence is also determined by the number of observed cases.

The four-field method can be applied not only for descriptive but also for quantitative indicators. What is characteristic is that the calculated coefficient K does not contain the actual initial data (values) but only their frequency. Therefore, the estimation of the correlation dependence which is acquired with the help of the coefficient for linear dependency, r_{xy}, is much more accurate.

Example

In a textile plant, a total of 1000 products have been manufactured from high quality materials, of which 990 are I grade and 10 are II grade. A further 1000 products have been manufactured from low quality materials, of which 510 are I grade and 493 are II grade (Table I.5.3). Determine the degree of the causal dependency between the quality of the materials and the quality (grade) of the manufactured products.

Table I.5.3. Input table for calculation of the degree of the causal dependency for the example

X \ Y	**I grade production**	**II grade production**	**Sum**
High quality material	990	10	1000
Low quality material	510	490	1000
Sum	1500	500	2000

(Continued)

(*Continued*)

The coefficient K is determined according to formula (I.5.6):

$$K = \frac{a \cdot d - b \cdot c}{\sqrt{(a+b)\cdot(c+d)\cdot(a+c)\cdot(b+d)}} = \frac{990 \times 490 - 10 \times 510}{\sqrt{1000 \times 1000 \times 1500 \times 500}} = +0.551.$$

The value of the coefficient proves that there is a direct (+ sign) and significant ($K > 0.5$) dependence between the grade of the production and the quality of the raw material.

MULTIPLE AND PARTIAL CORRELATION

In textile practice it is very often the case that the event of the effect is affected by several causes, that is, there is an interdependence between three or more random variables. For example, the strength up to the point of breakage of a yarn depends on the linear density of the fibers, their length, the twist of the yarn, and so on; the breakage in spinning and weaving depends on the speed of the spindles, respectively the looms, but also on other parameters, such as the irregularity of the yarn. In these cases the method of multiple correlation is used, which allows an assessment as to what extent the combined effect of several factors, such as the length and the linear density of the fibers affects the quality (strength up to the point of breakage) of the yarn produced. At the same time it is possible to assess which of the two combined factors has a greater impact and to what extent on the event of the effect (strength up to the point of breakage). This so-called correlation analysis is one of the most often used methods of mathematical statistics that has a wide application in regression analysis.

This chapter will review only the coefficient of double correlation and the partial coefficient of correlation in a case with three variables, and the general case of the multiple correlation coefficient. The relationship between the coefficient of correlation and the regression coefficients will also be reviewed in detail in the Regression analysis section.

Let us examine the case when we try to determine the influence of two causes x and y on the effect z. The *coefficient of double correlation* is determined with the help of the following formula:

$$R_{z(x,y)} = \sqrt{\frac{r_{xz}^2 + r_{yz}^2 - 2r_{xz}r_{yz}r_{xy}}{1 - r_{xy}^2}}, \tag{I.5.7}$$

where r_{xz}, r_{yz}, and r_{xy}, are the linear correlation coefficients, respectively, between x and z, y and z, and x and y. The coefficient of double correlation $R_{z(x,y)}$ is always positive and ranges between 0 and 1. When $R_{z(x,y)} = 0$, the variable z has no linear relationship with x and y. However, it is possible that there is some other (nonlinear) dependence. When $R_{z(x,y)} = 1$, there is a functional linear dependence between x, y, and z which is of the following type: $z = a \cdot x + b \cdot y + c$.

The *partial coefficient of correlation* is used when the aim is to establish the degree of dependence of the effect on one of the causes by neglecting the influence of the other cause. For example, what is the influence of the length of the yarn fibers on the strength up to the point of breakage if the other factor (the linear density of the fibers) does not change, or the opposite, the influence of the linear density of the fibers if the impact of the length is neglected? The partial coefficients of correlation in case there are three variables are determined with the following formulas:

$$r_{zx,y} = \frac{r_{zx} - r_{xy} \cdot r_{zy}}{\sqrt{(1-r_{zy}^2)(1-r_{xy}^2)}} \tag{I.5.8}$$

and

$$r_{zy,x} = \frac{r_{zy} - r_{xy} \cdot r_{zx}}{\sqrt{(1-r_{zx}^2)(1-r_{xy}^2)}}, \tag{I.5.9}$$

where $r_{zx,y}$ determines the degree of dependency between x and z when y is neglected and $r_{zy,x}$ is the degree of dependence between y and z when x is neglected. The partial coefficient of linear correlation has the same qualities as the classic coefficient of linear correlation r_{xy} and falls within the interval $[-1; +1]$.

Example

Assess the dependence of the strength up to the point of breakage on the length of the fibers and their linear density. The following values of the linear correlation coefficients have been established in preliminary studies: between the length of the fibers and the strength of the yarn, $r_{xz} = 0.8$, between the linear density of the fibers and the strength of the yarn, $r_{yz} = 0.85$, and between the length of the fibers and their linear density, $r_{xy} = 0.75$.

The coefficient of double correlation is determined with the help of formula (I.5.7):

$$R_{z(x,y)} = \sqrt{\frac{0.8^2 + 0.85^2 - 2 \times 0.8 \times 0.85 \times 0.75}{1 - 0.75^2}} = 0.885.$$

Its value shows that there is a strong linear dependence between the length of the fibers, their linear density, and the strength of the yarn they have been used for.

The partial coefficients of correlation are determined according to formulae (I.5.8) and (I.5.9):

$$r_{zx,y} = \frac{r_{zx} - r_{xy} \cdot r_{zy}}{\sqrt{(1-r_{zy}^2)(1-r_{xy}^2)}} = \frac{0.8 - 0.75 \times 0.85}{\sqrt{(1-0.85^2)(1-0.75^2)}} = 0.466,$$

$$r_{zy,x} = \frac{r_{zy} - r_{xy} \cdot r_{zx}}{\sqrt{(1-r_{zx}^2)(1-r_{xy}^2)}} = \frac{0.85 - 0.75 \times 0.8}{\sqrt{(1-0.8^2)(1-0.75^2)}} = 0.397.$$

The resulting values for the partial coefficients of correlation show the strong influence of the interaction between the length and the linear density of the fibers on the strength of the yarn.

I.6. Analysis of Variance

The analysis of variance (ANOVA) is one of the methods of mathematical statistics that has the widest field of application. It is used to study the influence of one or more factors on a certain variable. The typical problems which can be solved with the help of ANOVA in textile practice are related to establishing the influence of a certain additive, adjusting, or another technical parameter on the properties of the final product, for example, establishing the influence of the twist on the strength of the yarn.

SINGLE-FACTOR ANALYSIS OF VARIANCE

It is used to study the influence of the factor F on the property x. The factor, F, takes up k number of levels, marked with F_i, and for every level of the factor n number of observations are conducted. The individual values from the measurements, x_{ij}, and the mean values for every level of the factor, $\overline{x}_i$, are entered in a table, the outline of which is given in Table I.6.1.

The mean value, $\overline{\overline{x}}$, of all tests is determined on the basis of the individual mean values, $\overline{x}_i$:

$$\overline{\overline{x}} = \frac{1}{k}\sum_{i=1}^{k} \overline{x}_i. \tag{I.6.1}$$

The sum total of the squares, Q, is determined:

$$Q = \sum_{i=1}^{k}\sum_{j=1}^{n}\left(x_{ij} - \overline{\overline{x}}\right)^2, \tag{I.6.2}$$

which is expanded into two components:

$$Q = Q_F + Q_e, \tag{I.6.3}$$

where Q_F is the sum of the squares of the deviations between the groups of tests, Q_e is the sum of the squares of the deviation inside the groups (residual sum).

The sum Q_F is determined as follows:

$$Q_F = n\sum_{i=1}^{k}\left(\overline{x}_i - \overline{\overline{x}}\right)^2 \tag{I.6.4}$$

Table I.6.1. Input table for single-factor ANOVA

Factor \ Test	1	2	3	...	n	$\bar{x}_i$
F_1	x_{11}	x_{12}	x_{13}	...	x_{1n}	$\bar{x}_1$
F_2	x_{21}	x_{22}	x_{23}	...	x_{2n}	$\bar{x}_2$
...	...	...	...	...	...	...
F_k	x_{k1}	x_{k2}	x_{k3}	...	x_{kn}	$\bar{x}_k$

and is used to assess the influence of the factor, as it is determined by the deviations of the mean values for the individual levels of the factor and the common mean value. The sum Q_e is calculated from the values of the individual tests and the mean values for the respective levels:

$$Q_e = \sum_{i=1}^{k}\sum_{j=1}^{n}(x_{ij} - \bar{x})^2. \qquad (I.6.5)$$

It characterizes the influence of the random errors.

From the sums estimated in that way (between the groups, in the groups, and the total sums) the variances S_F^2, S_e^2, and S^2 are determined (Table I.6.2), taking into account the degrees of freedom.

With their help, through Fisher's criterion, the influence of the factor on the studies property can be assessed:

$$F_R = \frac{S_F^2}{S_e^2}. \qquad (I.6.6)$$

If $F_R > F_T$, it is accepted that the factor has an influence on the studied property and if $F_R \leq F_T$, it does not influence it.

Table I.6.2. Formulae for single-factor ANOVA in tabular form

Variance components	Sum of the squares	Degrees of freedom	Variances	F_R	F_T
Between groups	Q_F	$f_1 = k - 1$	$S_F^2 = \frac{Q_F}{k-1}$	$\frac{S_F^2}{S_e^2}$	f_1, f_2, α
In groups	Q_e	$f_2 = k\,(n - 1)$	$S_e^2 = \frac{Q_e}{k(n-1)}$		
Total	Q	$f = k \cdot n - 1$	$S^2 = \frac{Q}{k \cdot n - 1}$		

Example

A number of yarns have been spun from wool/rayon blend with different proportions of the two components in order to study the influence of the factor "rayon content" on the yarn strength. Six different variants have been manufactured and for each variant a total of eight measurements of the yarn strength have been made (in cN). The results have been summarized in Table I.6.3—columns (1–9). Determine whether the change in the proportion of the components has an influence on the yarn strength.

Table I.6.3. Measured values of the yarn strength

Measurement number / Proportion	1	2	3	4	5	6	7	8	$\bar{x}_i$	S_i^2
(1)	**(2)**	**(3)**	**(4)**	**(5)**	**(6)**	**(7)**	**(8)**	**(9)**	**(10)**	**(11)**
40:60	100	141	147	126	133	101	148	128	128	353.14
45:55	93	120	123	105	109	103	130	143	115.75	265.36
50:50	101	87	88	105	93	102	73	94	92.875	107.27
55:45	127	83	80	87	96	109	68	97	93.375	337.98
60:40	111	84	83	96	83	90	101	64	89	200.00
65:35	69	86	80	82	80	99	81	71	81	85.14

Column (10) shows the calculated mean values for the different variants (groups) and column (12) shows the variances in the groups. The common mean value is $\bar{\bar{x}} = 99.33$ cN. The different components of the variance are determined according to formulae (I.6.4) and (I.6.5) and the accountable value of Fisher's criterion is determined according to formula (I.6.6). The resulting values are given in Table I.6.4.

Table I.6.4. Calculation of the variance components for the example

Variance components	Sum of the squares	Degrees of freedom	Variances	F_R	F_T
Between groups	12869.75	$f_1 = 6 - 1 = 5$	2573.95	11.45	2.44
In groups	9442.25	$f_2 = 6 \times (8 - 1) = 42$	224.82		
Total	22312	$f = 6 \times 8 - 1 = 47$	474.72		

As $F_R = 11.45 > 2.44 = F_T\,(f_1 = 5, f_2 = 42, \alpha = 0.05)$, the differences in the mean values for the individual groups are significant and, therefore, the studied factor "rayon content" influences the strength of the studied yarn.

MULTIFACTOR ANOVA

The multifactor analysis of variances is used when the influence of more than one factor is studied. The *two-factor ANOVA* has the widest scope of application. It determines the influence of two factors A and B on the studied indicator. The number of levels for factor A is marked with l, and k marks the number of levels for factor B. For every combination between the levels of the factors A and B are conducted n number of measurements and the mean values are recorded in Table I.6.5.

Table I.6.5. Input table for multifactor ANOVA

B \ A	A_1	A_2	...	A_l	$\overline{x}_{i0}$
B_1	$\overline{x}_{11}$	$\overline{x}_{12}$	...	$\overline{x}_{1l}$	$\overline{x}_{10}$
B_2	$\overline{x}_{21}$	$\overline{x}_{22}$	...	$\overline{x}_{2l}$	$\overline{x}_{20}$
...	...	...	...	...	...
B_k	$\overline{x}_{k1}$	$\overline{x}_{k2}$	...	$\overline{x}_{kl}$	$\overline{x}_{k0}$
$\overline{x}_{0j}$	$\overline{x}_{01}$	$\overline{x}_{02}$	...	$\overline{x}_{0l}$	$\overline{\overline{x}}$

Table I.6.6. Formulae for multifactor ANOVA in tabular form

Variance components	Sum of the squares	Degrees of freedom	Variances	F_R	F_T
Between groups forfactor A	$Q_A = n \cdot k \sum_{j=1}^{l} \left(\overline{x}_{0j} - \overline{\overline{x}}\right)^2$	$f_A = l - 1$	$S_A^2 = \dfrac{Q_A}{l-1}$	$\dfrac{S_A^2}{S_e^2}$	f_A, f_e, α
Between groups for factor B	$Q_B = n \cdot l \sum_{i=1}^{k} \left(\overline{x}_{i0} - \overline{\overline{x}}\right)^2$	$f_B = k - 1$	$S_F^2 = \dfrac{Q_B}{k-1}$	$\dfrac{S_B^2}{S_e^2}$	f_B, f_e, α
From the interaction	$Q_{AB} = n \sum_{i-1}^{k} \sum_{j=1}^{l} \left(\overline{x}_{ij} - \overline{x}_{i0} - \overline{x}_{0j} + \overline{\overline{x}}\right)^2$	$f_{AB} = (k-1)(l-1)$	$S_{AB}^2 = \dfrac{Q_{AB}}{f_{AB}}$	$\dfrac{S_{AB}^2}{S_e^2}$	f_{AB}, f_e, α
Residual	$Q_e = \sum_{i=1}^{k} \sum_{j=1}^{l} \sum^{n} \left(x_{ij} - \overline{x}_{ij}\right)^2$	$f_e = k \cdot l(n-1)$	$S_R^2 = \dfrac{Q_e}{f_e}$		
Total	$Q = \sum_{i=1}^{k} \sum_{j=1}^{l} \sum^{n} \left(x_{ij} - \overline{\overline{x}}\right)^2$	$f = k \cdot l\, n - 1$	$S^2 = \dfrac{Q}{f}$		

The mean values for the columns are determined (factor A) $\overline{x}_{0j}$, for the rows (factor B) $\overline{x}_{i0}$, and the total mean value $\overline{\overline{x}}$.

In a similar way to the single-factor ANOVA, between the groups, in the groups (residual), and the total variances are determined—the variance between the groups being calculated separately for factor A and factor B. With the help of the two-factor ANOVA, the influence and the interaction between the two factors can be assessed. For this purpose a sum, respectively, a variance of the interaction should be defined. The calculation process for the sums, the degrees of freedom, and the variances is shown in Table I.6.6.

To assess the influence of the individual factors and the interaction of the two factors on the studied property, the estimate values of Fisher's criterion F_R are used. The way they are determined and the conditions used to account for the table values F_T are given in the last two columns of Table I.6.6.

Example

Study the influence of the needle number and the density of the seam on the strength of the seam (in N). The study has been conducted with three different needle numbers: 70, 80, and 90, the density of the seam being 3, 4, and 5 stitches/cm. Five tests have been conducted for each combination of the factors. The data from the measurements is given in Table I.6.7.

Table I.6.7. Measurement data

Density \ No	70	80	90
3	192.9	155.3	183.5
	220.9	170.3	188.5
	199.1	168.2	184.5
	195.6	160.5	174.5
	176.0	176.6	185.2
4	243.6	230.1	252.7
	254.0	201.5	210.0
	261.1	224.9	218.0
	255.7	230.1	245.7
	224.1	209.9	233.4
5	267.7	218.1	229.7
	277.2	250.4	253.2
	264.4	220.4	203.5
	293.3	156.0	247.1
	256.3	218.3	271.6

The mean values for each factor combination, under columns, rows, and the total mean value are given in Table I.6.8.

Table I.6.8. Calculated mean values for the example

Density \ No	70	80	90	$\bar{x}_{i0}$
3	196.90	166.18	183.24	182.11
4	247.70	219.30	231.96	232.99
5	271.78	212.64	241.02	241.81
$\bar{x}_{0j}$	238.79	199.37	218.74	218.96

The calculation of the sums, the degrees of freedom, and the variances is done accordingly in Table I.6.6, and the results are presented in Table I.6.9.

Table I.6.9. Sums of the squares, degrees of freedom, and variance components for the example

Variance components	Sum of the squares	Degrees of freedom	Variance	F_R	F_T
Between groups for factor *A*	31158.41	$f_A = 3 - 1 = 2$	15579.21	44.99	3.259
Between groups for factor *B*	11656.49	$f_B = 3 - 1 = 2$	5828.25	16.83	3.259
From the interaction	1485.90	$f_{AB} = (3 - 1) \times (3 - 1) = 4$	371.47	1.072	2.634
Residual	12465.54	$f_e = 3 \times 3 \times (5 - 1) = 36$	346.26	—	—
Total	56766.34	$f = 3 \times 3 \times 5 - 1 = 44$	1290.14	—	—

As can be seen from the resulting values for the estimate and table coefficients of Fisher, both factors influence the strength of the seam ($F_R > F_T$). The influence of the density is stronger (factor *A*) since the accountable value of the Fisher's criterion is bigger than the value of the needle size (factor *B*). The interaction (the joint effect) of the two factors is insignificant since ($F_R = 1.072 < 2.634 = F_T$).

PART II

Design of an Experiment

In contemporary science the experiment has a central place. The efficiency of scientific investigations increases through the use of mathematical methods.

It was in the beginning of 1935 when the English statistician Ronald Fisher advanced the first statistical methods for the design of an experiment. He proves the advisability of the simultaneous variance of all factors instead of the then widely used monofactorial experiment.

The *design of an experiment* is a process of a choice of the number and conditions for conducting of the tests, necessary and sufficient for solving the given task with the necessary accuracy. In 1951, Box and Wilson introduced a new direction in the design of an experiment related to optimization of the processes. Finding the optimal conditions for implementation of the processes is one of the main scientific and production tasks.

An experiment conducted for solving of an optimization task is called an *extreme experiment.*

II.1. Main Concepts in Mathematical Modeling and Optimization

Mathematical modeling is done on objects. "An object" refers to all technical means and phenomena. The aim of the researcher is to study the objects so that he can manage them and ensure the best (optimal) accomplishment of the processes. Each production object is characterized by a number of indices of its working $y_1, y_2, \ldots, y_n$ (Figure II.1.1).

They are called *output variables of the object*, *target functions*, or *parameters for optimization*. The values the output variables take depend on a number of other variables. Those of them that can be modified, or at least measured, are called *input variables* or *factors* $(x_1, x_2, .., x_m)$.

A number of other variables $(w_1, w_2, \ldots, w_p)$ that cannot be modified usually also act on the objects. They are called *random disturbances*. Often the physical nature of the random disturbances is not known and only their action is noticed in measuring the output variables. As a result of the action of the random disturbances, different values of the output variables can be obtained with one and the same value of the factors. For example, if an object of the study is yarn, the *input parameters* are linear density of the fibers, strength of the fibers, friction coefficient between fibers, twist, and many other material and machine parameters. The *output parameters* are strength, elasticity, unevenness of the cross section, hairiness, and so on. *Random disturbances* are change in temperature, humidity of the environment, uneven drawing of fibers by the drawing devices, and so on.

For study of the objects, *mathematical models* are used. A mathematical model means a mathematical correlation determining the quantitative relation between the output variables and the factors. Mathematical models reflect separate characteristics of the object that the researcher is interested in. In addition, they reproduce these characteristics with a different degree of accuracy. This is the reason why for one and the same object, a few models can exist reflecting different aspects of its work and describe it with different accuracy. Mathematical models are classified according to a number of principles. One of them is the way the models arc formed. According to this indication, they can be *analytical*, *experimental*, and *analytically experimental*.

The main physical and chemical regularities for an object are used in the formation of *analytical models*. The models themselves, most of the time, are systems of algebraic and differential equations (in ordinary or partial derivatives). A very good knowledge of the mechanism of the processes taking place within the object is necessary for their formation.

Experimental models are obtained on the basis of data from the measurement of the input and output quantities of the object. No information for the mechanism of the phenomena is

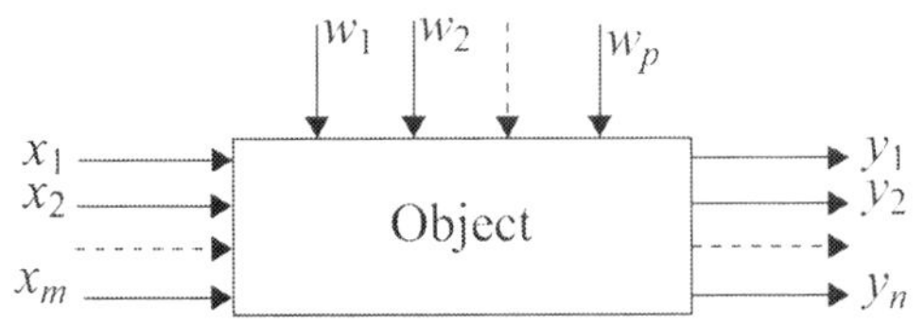

Figure II.1.1. Scheme of the influences on an object.

necessary for them to be formed although it turns out useful in this case, too. That is why experimental models are especially appropriate for examination of objects with insufficiently clarified nature of the phenomena taking place in them. Unfortunately, there still are many such phenomena and objects in textile practice.

Analytically experimental methods combine aspects of both approaches. In this case, the structure of the models is chosen mainly for physical and chemical reasons but a number of constants are determined experimentally.

Models can also be classified according to whether they describe the statics or dynamics of objects. *Models of dynamics* reflect the development of processes in time while *models of statics* describe established regimes. In determination of the optimal conditions for the working of a given object, mainly the models of statics are an object of interest and for this reason they are the ones to be examined. Dynamic models are of interest mainly to the specialist of automatic management and also for some special technological investigations such as carding, tension of threads in winding and weaving, and in finishing.

A number of other indications exist according to which models can be classified. For example, dynamic models may be with *concentrated parameters* in which factors are modified only with time or with *distributed parameters* when the model depends on a few parameters.

An *experiment* may be *passive* and *active.* In *passive experiment*, information for the object is obtained during its normal exploitation, that is, without modification of the values of the factors. Result processing is done by a regression and correlation analysis. The method of consecutive exclusion of constituting functions (method of Brayton) can be used. Correlation, spectral or dynamic regression analysis, is used to obtain a dynamic model. In *active experiment*, levels of input data are artificially modified. The aim is for the reliability of the results to be increased.

Experimental research aims to determine the dependency of the output variable on the factors on the basis of an experiment data. The approach used by researchers until the appearance of the theory of experiment was the *classical* or *traditional* approach. In this approach, the investigator fixes $(m - 1)$ factors on constant values and then modifies the other factor and measures the result y_i. Then, he fixes the value of this factor and starts modifying another one. This goes on until the influence of all factors on the output variables is studied. Results are presented by tables and graphs according to which the technological regime is chosen.

In order to obtain a full enough picture of the work of the object, a few (e.g., q) values need to be given for each of the factors. The number of tests in which all possible combinations of q values of the m factor are realized is $N = q^m$. For example, if the object is six-factorial and each of the factors is given 5 values, in order for a full enough research of the object to be made with the traditional approach, the number of tests necessary is $N = 5^6 = 15{,}625$ tests. As conducting so many tests is labor-intensive and often very expensive, the experimenter has to choose (on the basis of his experience and intuition) constant values of the factors and to conduct multiple monofactorial experiments. However, they do not allow for the influence of different combinations of factors on the results to be seen and there is no guarantee that the chosen technological regime is optimal.

The approach of the theory of experiment differs considerably from the traditional one. The first difference is that it uses mathematical models. Determination of the model is possible on the grounds of a relatively small number of tests. If they are conducted correctly, the model gives the possibility for the value of the output variable to be predicted with sufficient accuracy in random technological regimes from the field of modification of the factors. In the traditional approach, reliable information is obtained only for the regimes in which observations have been made. Using methods for optimization, the best regime for given indications can be found with the help of computing technology. The results obtained are guaranteed with preliminary set probability.

The theory of experiment solves the issue regarding planning of the tests. Technological regimes exist in each specific experiment and by studying them we can extract maximum information for the studies object. The list of these regimes is the optimal plan of the experiment. *The choice of an optimal plan of the experiment and ensuring maximum accuracy of prediction of the output variable are the big advantages of the theory of experiment.*

Unlike the traditional approach, the theory of experiment recommends that all factors are modified simultaneously in all tests. For this reason, averaging of the results from all tests is done during the processing of the data for determination of the constants. The random error is filtered and the accuracy of the models is increased. The opportunity for an optimal technological regime to be obtained with only a few tests leads to:

- increase in the quantity of the production with improvement or maintenance of its quality;
- preservation or increase in the quantity and quality of production with substituting expensive raw materials with cheaper or easily available ones;
- increase in productivity;
- shortening of the periods for investigating and accelerated implementation of new technologies; and
- decrease in the number of tests and the executive staff.

II.2. Choice of Parameters of Optimization

In modeling and optimization of multifactorial objects, the researcher does not take into account many aspects of the studied object but focuses his attention on only some of them. He sets a specific target and in order to achieve it, characterizes it in a clear and accurate manner. The characteristic of the research target is called *a target function, a parameter of the modeling,* and *optimization* or *an output parameter, y* (Figure II.1.1). The real multifactorial objects are exceptionally diverse and for this reason there are various target functions which can be chosen for the different objects, or even for one and the same object.

The target functions could be the following:

- economic (productivity, profitability, cost price, profit, and so on)
- technical–economic (coefficient of efficiency, security, durability, and so on)
- technological (physical, chemical, biological, mechanical, and other characteristics of the production or the product)
- psychological
- aesthetic, etc.

The diversity of the possible target function is huge. The choice of one of them is directly related to the specific object and the targets of the researcher. The choice of a target function is a basic stage of the modeling and optimization process and, therefore, it should guarantee the possibility for the application of different statistical methods.

The target function should meet the following requirements:

1. The target function should be quantitatively characterized, that is, it should be possible to express it with a number. The multitude of values a target function can obtain is called the area of its determination. This area can be limited or unlimited, continuous or discrete. In order for a target function to be expressed in a numerical manner, it (or the components used to calculate it) should be measurable. In those cases when there are no appropriate measuring devices or there is an existing possibility to measure the output parameter, the so-called ranking approach is used. The rank is a conditional (subjective) quantitative evaluation of the target function on a preselected scale. In the simplest case, a ranking scale has only two values marked with two numbers, for example, 0 and 1. These are cases when the output is evaluated only as fit or defective, of high or of poor quality, and so on. In other cases the ranking scale might have more discrete values given. A typical example of this is the scale for color fastness.

It is obvious that the ranking approach is less accurate and too subjective. This is the reason why it is used only when there is no possibility for the target function to be measured in an objective manner.

Obtaining a quantitative characteristic of a target function is a necessary requirement in modeling and optimization since this is the only way to apply mathematical statistical methods.

2. The target function should be a single function. In many cases this requirement is met as it comes directly from the research target. But in other cases the target is given in a more general manner and several output parameters can be outlined. For example, in the production of fabrics different targets can be set, for example, high strength, elasticity, low shrinkage, abrasion, and washing resistance. It is very difficult to give a preference to only one of these parameters. The quality of the product is determined as a totality of the specified targets.

   When we are interested in each output parameter, y_j, separately (Figure II.1.1), the modeling and optimization of this parameter are done in compliance with the methods reviewed above. But when the target of the research concerns all n output parameters, there are three ways of choosing a target function y:

- A generalized target function is formulated from all output parameters:

$$y = \varphi(y_1, y_2, ..., y_n); \qquad \text{(II.2.1)}$$

- A generalized target function is formulated for some u ($u < n$) output parameters:

$$y = \varphi(y_1, y_2, ..., y_u), \qquad \text{(II.2.2)}$$

  while the rest ($u - n$) parameters are examined as additional restrictions;
- Only one of the parameters y_1, y_2, …, y_n is chosen as a target function and all the others are examined as restrictions.

The formulation of a generalized target function of the type (II.2.1) or (II.2.2) is a complex task. The generalized target function set in the following way is easier to formulate and rationalize:

$$y = \sum_{j=1}^{u} a_j \cdot y_j, \qquad u = 2, 3, \ldots, n, \qquad \text{(II.2.3)}$$

where α_j are weight coefficients that take into account the influence of the particular parameters in the quantitative characterization of the target.

Weight coefficients could be positive (when the increase of the respective parameter corresponds to the target) or negative (when the increase of the parameter leads to deviation of the target set) quantities. These coefficients are set by the researcher and in most cases have a subjective character. They may also be determined as ranks, taking into consideration the opinion of a group of experts.

Since the separate parameters, $y_1, y_2, \ldots, y_n$, have different dimensions and variation intervals, the coefficients α_j should also have different dimensions and variation intervals.

That is why it is easier to proceed with the formulation of a target function as a dimensionless variable limited only to the (–1, +1) interval. With this objective, the parameters $y_1, y_2, \ldots, y_n$ are transformed into new parameters $\theta_1, \theta_2, \ldots, \theta_n$ with the help of the following formula:

$$\theta_j = \frac{y_j - \bar{y}_j}{\frac{y_{j,\max} - y_{j,\min}}{2} \cdot u}, \qquad j = 1, 2, \ldots, u \leq n, \tag{II.2.4}$$

where $\bar{y}_j$ is the mean value of y_j, calculated as follows:

$$\bar{y}_j = \frac{y_{j,\max} - y_{j,\min}}{2}; \tag{II.2.5}$$

where $y_{j,\min}$ and $y_{j,\max}$, respectively, are the minimum and the maximum value of y_i,
u is the number of parameters included in the generalized target functions.

After the transformation, formula (II.2.3) becomes

$$y = \sum_{i=1}^{u} a_j \cdot \theta_j. \tag{II.2.6}$$

In this case, if the coefficients a_j are set as dimensionless variables limited to the (–1, +1) interval, the generalized target function y, determined with formula (II.2.6), would also be dimensionless with values limited to the same interval. The choice of a target function in this way has several advantages in terms of calculation and in comparing the influence of the individual parameters. The inconvenience of this choice is that the direct physical meaning of the target function is lost.

In the usage of generalized target functions, determined according to the following formulae: (II.2.1), (II.2.2), or (II.2.6), there are two possible approaches to modeling and optimization of the object. The first approach is to build a model of the generalized output parameter formulated in this way that would be tested and optimized further. This approach is advisable in the beginning of the modeling and the optimization process. The second approach is related to building separate models for each parameter, y_i, and additional formulation, with formula (II.2.6), of the generalized target function. This approach requires more work (building the u model) but may appear to be more appropriate when the coefficients a_j have to be modified or when the structure of the generalized target function has to be modified.

3. The target function has to be efficient, that is, it has to characterize in an effective manner the functioning of the studied object in compliance with the set target. This requirement determines the accuracy of the formulation of the task for modeling and optimization. The efficiency of the target function does not remain constant in the process of studying complex objects. It changes according to the achieved results. This often leads to the use of a consecutive approach in the choice of the output parameter. For example, in the initial stage of a research, the amount of production per unit of time may be chosen as a target function. However, when the final capabilities for productivity increase are reached, the role of other parameters such as the quality or the cost price of the product grows significantly.

At the same time, the target function has to be as universal as possible, that is, it should characterize in a comprehensive and complete way the target of the research. The more universal it is, the less often it has to be changed. In this sense, the economic parameters are more universal than the technical ones. The generalized target functions in principle have a higher level of universality compared to the separate output parameters.

4. The target function should be a statistically effective characteristic of the target, that is, it should have the minimum possible dispersion. In practice, this requirement comes down to choosing such a target function which can be determined (measured or calculated) with the highest accuracy.

There may be many more requirements to the target functions. The requirement for a target function to have a clear physical meaning is of great importance for the interpretation of the results. The direct output parameters of the object (e.g., productivity, strength) have that property while the transformed or the generalized target functions do not have specific meaning, which makes the interpretation of the models and results of the optimization more difficult.

Another requirement related to the practical implementation of the modeling and optimization process is that the target function should be as simple as possible and easy to calculate.

All these requirements show that the choice of a target function is not a routine task. The researcher has to be careful with it as no objective method for choice of a target function exists. Now is the moment to mention that economic parameters, due to their additive nature, are approximated more easily with linear models while the physical and chemical parameters are more often described with more complex functions and a linear approximation is possible only in narrower intervals of factors variation.

II.3. Choice of Input Factors

Input factors are variables which can have certain values and can be measured. In order for the mathematical model of the examined multifactorial object to be adequate and have a practical application, the most important and essential factors which determine its functioning have to be included. At the same time, as few factors and their functions as possible should be included in order for the model to be convenient for use. *Therefore, the task for the choice of the factors in the modeling is in fact the determination of the necessary requirements, the choice of the essential factors, and the elimination (screening) of the nonessential factors.*

In the preliminary examination of the object, a definite number of factors are selected. As the omission of an essential factor leads to obtaining inadequate models, the researcher always includes in the list many more factors than necessary. The factors used for the construction of the model have to meet (separately and in totality) several basic requirements. Each separate factor x_i should have a certain interval (area) of determination $(x_{i,\min}, x_{i,\max})$ in which it should allow for a quantitative evaluation, that is, it should be *controllable*. The evaluation could be obtained thorough measurement or a rank nature. However, in both cases, the evaluation of the factor should be definite.

In addition to the requirement of controllability, when conducting an active experiment the factors should also be *manageable*. This means the existence of a possibility for the factors to be determined and maintained with sufficient accuracy on different levels in the area of their determination. Such a requirement cannot always be met and this often is the main impediment for conducting an active experiment. The measurement of the factors should be done with *the highest possible degree of accuracy*. That is why when two or more factors can be measured in several ways, or one and the same physical variable may be expressed (directly or indirectly) through several factors, the method or the variable that ensures the highest degree of accuracy has to be chosen. The inaccuracies of the measurement may lead to obtaining shifted evaluations of the coefficients of the model and to reduction of the coefficient of multiple correlation. It has to be taken into consideration that the factors which have a narrow variation interval and which appear to be the most essential should be measured with the highest accuracy possible.

Together with these main requirements for each separate factor, the population of factors should also provide for the implementation of two additional requirements: *independence* and *compatibility of the factors*. Independence means that in the population there are not two functionally related factors. In an active experiment, this requirement allows for every factor to be determined on a certain level, regardless of the levels of the other factors. In a passive experiment it ensures relatively low correlation between the factors. In both types of experiments, independence of the factors ensures obtaining additional information from each factor. The requirement for compatibility of the factors ensures the possibility of implementation for all possible combinations of values of the factors within the limits of their areas of determination. In real cases, this requirement cannot always be met. However, the area of compatibility needs to be studied and determined very well. This is especially important in the conducting of an active experiment, as in passive experiments the regime of normal exploitation ensures the implementation of the compatibility requirement. If this requirement is not implemented, it means that there are combinations of factor values leading to waste, to destruction of the object, or to other unacceptable events.

The next issue is the *choice of factors* through determination of the essential ones and screening of the nonessential. The formulation of the task is the following: each factor influences the output parameter in a certain way and this influence is different for different factors. If the factors are arranged (ranked) by the extent of influence on the output parameter, we will get a picture similar to the one in Figure II.3.1. In this figure, the influence of the factors is represented in percentage on the y-axis while the factors in the order of their influence are shown on the x-axis. The total (cumulative) effect of the activity of the separate factors is also shown.

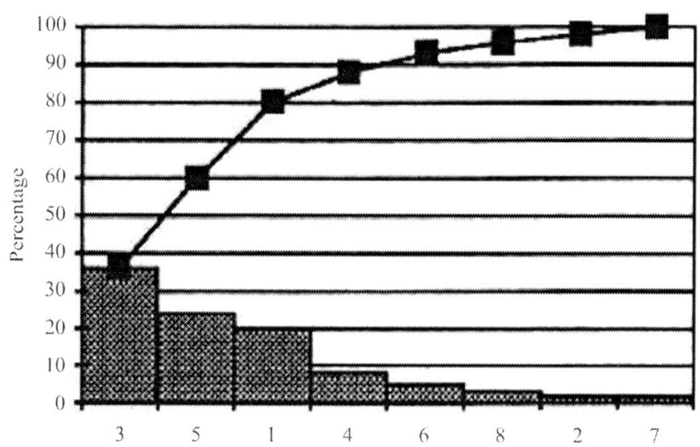

Figure II.3.1. Factors ranked by the extent of influence on the output parameter.

Obtaining such ranking of the factors practically solves the problem of choosing the essential factors and screening the nonessential ones. Unfortunately, this is a very hard and insufficiently formalized problem. In practice, it is solved consecutively by starting at the stage of the preliminary tests and ending with the statistical analysis of the model.

The whole process of choosing the essential factors may be summarized in the following stages:

1. Selection of the essential factors on the basis of research of the object nature: Usually the researcher has significant a priori information that often allows him/her to choose the essential factors, or at least the main part of them. However, in many cases the available information is not sufficient for a definite choice and serves only to focus on certain factors. Also, there is a possibility of a case when the object is so complex or poorly studied that the a priori information does not provide the possibility to make a final decision. Then you need to proceed to the next stages.

2. Selection of essential factors through a priori ranking: When the information about the object possessed by the researcher is not sufficient for making an informed choice of the essential factors, the opinion of a group of experts with a lot of experience in the field has to be taken. An inquiry is organized in which each expert is asked to prepare a list of the essential factors arranged to the extent of influence on the output parameter. The prevailing opinion of the experts is used for the choice of the essential factors. Experience shows that to a great extent this way of a priori ranking gives positive results. For higher reliability of the inquiry, greater number of different experts should participate. In some cases the opinions of the experts may be accounted for with different weight coefficients depending on their qualification. The methods of the ranking correlation are presented later in this chapter.
3. When the opinions of the experts are divergent on many of the factors, the only option left is to conduct a small experiment and then select the factors on those grounds. This is the stage of the preliminary experimental screening of the factors. The main concept of the experimental screening is based on the suggestion that the most essential factors can be selected even when there is little experimental data. For this purpose, a number of mathematical methods are used such as dispersion analysis (presented in Section I.6) and the random balance method (presented later in this chapter). With this stage, the possibilities for preliminary screening of essential factors are exhausted. The researcher forms the final list of factors and proceeds to the next stages of modeling. If not all essential factors are included in the model (i.e., the model is found to be inadequate), the researcher would need to return to the stage of factor selection once again and to select new factors.

As it can be seen, the selection of factors is a long and difficult process that ends with the stage of the statistical analysis of the model.

In the active factorial experiment, factors most often vary on two levels. Before the levels of the factors are determined, the center/core of the experiment is chosen. A *center of the experiment* is a point in the factorial space, the coordinates of which are the zero (central) levels around which the separate factors vary. It is desirable that the center is as close as possible to the optimal value wanted. This gives the possibility to reach the final result with a minimum number of tests. The coordinates of the center, or the so-called *basic levels of the factors*, are marked with $X_{0,i}$.

After determining the center of the experiment, the next step is to choose the interval of variation, J_i, of the factors. In the choice of the interval of variation, the following considerations are taken in account:

- The values of the factor in the interval of variation should be measured with sufficient accuracy.
- The available information of the variation range of the output parameter in the different points of the factorial space.
- A reference for the size of the interval can be obtained if there is preliminary information for the area of response.

An *area of response* is the area in the $(m + 1)$-dimensional space, the points of which are determined by the values of the factors and the output variable of a particular test. It is described with the mathematical model of the object.

Depending on the minimum and the maximum values of the separate factors, the interval may have different width.

- In wide interval $J_i \geq 0.3(X_{i,\max} - X_{i,\min})$.
- In medium interval $0.1(X_{i,\max} - X_{i,\min}) \geq J_i \geq 0.3(X_{i,\max} - X_{i,\min})$.
- In narrow interval $J_i \leq 0.1(X_{i,\max} - X_{i,\min})$.

The maximum and the minimum values of the factors are determined by the technical conditions of the process implementation.

After the interval of variation is chosen, the upper and the lower levels of the factors are determined:

$$X_{i,u} = X_{0,i} + J_i, \tag{II.3.1}$$

$$X_{i,l} = X_{0,i} - J_i. \tag{II.3.2}$$

When writing the conditions of the experiment and calculation of the models, it is more convenient for the variables to be given in coded values instead of natural ones. Coding is performed in the following way:

$$X_i = \frac{\tilde{X}_i - \tilde{X}_{0,i}}{J_i}, \tag{II.3.3}$$

where X_i is the coded value of the factor,
$\tilde{X}_i$ is natural value of the factor,
$\tilde{X}_{0,i}$ is natural value of the basic level,
J_i is interval of variation,
i is number of the factor.

In this way, the interval of variation is [–1, +1] and code 0 corresponds to the basic level.

Example
The values should be coded in the conducting of an experiment with a drawing frame with the following variation of the speed of delivery: $\tilde{X}_u = 800$ m/min and $\tilde{X}_l = 600$ m/min.
Determination of the basic level in natural values:
$$\tilde{X}_0 = \frac{800 + 600}{2} = 700 \text{ m/min}.$$

(Continued)

Determination of the interval of variation:

$$J = \frac{800-600}{2} = 100 \text{ m/min}.$$

Determination of the coded values:

For the upper level: $X_u = \dfrac{800-700}{100} = +1,$

For the lower level: $X_l = \dfrac{600-700}{100} = -1,$

For the basic level: $X_0 = \dfrac{700-700}{100} = 0.$

METHODS OF RANK CORRELATION

Rank correlation is a method in mathematical statistics for objective processing and formalization of subjectively collected information designed for quantitative evaluation of objectively immeasurable factors and output parameters. The approach for solving such tasks is collection and summary of the opinions of a certain number of experts (specialists) as from the prevailing opinion and the final decision is made. The experts, who should be as many as possible, have essential but not formalized knowledge. For this reason, the collected information reflects the objective aspect of the phenomenon but also the subjective attitude of each expert, which is the main disadvantage of this approach. The collection of opinions of highly qualified experts has its advantages too. The knowledge, the experience, and the intuition of many experts are concentrated and summarized in the opinions, which makes this approach exceptionally useful.

The method is based on the "rank" concept, that is, the evaluations of the inquired experts are presented as ranks for each object. If n objects (phenomena, factors, parameters) of a given population are arranged (ranked) in accordance with the decrease or increase of a certain indicator, x, of theirs then the rank of each ith object, x_i, is a number showing the place of the object in the ordered population.

Example

The influence of constructive and technological parameters of the transport devices of a sewing machine on the set up of the stitch line should be ranked. The results of the expert in machines and devices in sewing production are summarized in Table II.3.1.

(Continued)

(*Continued*)

Table II.3.1. Ranking from the experts

Code of the factor	Factors	Rank
1	Material of the working device	7
2	Incorrect choice of the height of the feed dogs' teeth (in accordance with the type of the material)	5
3	Height of raising of the feed dogs over the needle plate	2
4	Parallelism of the teeth's and needle plate's surfaces	4
5	Pressure power of the presser foot	1
6	Area of the presser foot	6
7	Angle of the teeth of the feed dogs	8
8	Transportation speed	3
9	Number of rows of the feed dogs	9
10	Height of the teeth of the feed dogs	10

COEFFICIENT OF RANK CORRELATION

The coefficient of rank correlation ρ serves for assessment of the relation between two indicators x and y that cannot be quantitatively evaluated. If n objects are ranked in accordance with the indicators x and y, they are represented as $x_1, x_2, \ldots, x_n$, and $y_1, y_2, \ldots, y_n$.

The Spearman's coefficient of rank correlation is determined by the formula:

$$\rho = \frac{\sum_{i=1}^{n}\sum_{j=1}^{n}(x_j - x_i)(y_j - y_i)}{\sqrt{\sum_{i=1}^{n}\sum_{j=1}^{n}(x_j - x_i)^2 \cdot \sum_{i=1}^{n}\sum_{j=1}^{n}(y_j - y_i)^2}}, \qquad \text{(II.3.4)}$$

where x_i is the rank of the ith object according to the indicator x,
x_j is the rank of the jth object according to the indicator x,
y_i is the rank of the ith object according to the indicator y,
y_j is the rank of the jth object according to the indicator y.

The coefficient of rank correlation can also be determined according to Kendall:

$$\tau = \frac{\sum_{i=1}^{n}\sum_{j=1}^{n}\text{sign}(x_j - x_i)\cdot\text{sign}(y_j - y_i)}{\sqrt{\sum_{i=1}^{n}\sum_{j=1}^{n}\left[\text{sign}(x_j - x_i)\right]^2 \cdot \sum_{i=1}^{n}\sum_{j=1}^{n}\left[\text{sign}(y_j - y_i)\right]^2}}, \qquad \text{(II.3.5)}$$

where the symbol sign($a - b$) is determined in the following way:

$$\text{sign}(a-b) = \begin{cases} 0, & a = b, \\ +1, & a < b, \\ -1, & a > b. \end{cases} \quad \text{(II.3.6)}$$

Kendall's coefficient of rank correlation is a less effective evaluation than Spearman's coefficient. In addition, more calculations are needed for it to be calculated. That's why its use is unsuitable for, both, theoretical and practical reasons.

Taking into account that the ranks of all objects according to x and y take the whole values from 0 to n and the margins $(x_i - x_j)$ and $(y_i - y_j)$ take the whole values from $(1 - n)$ to $(n - 1)$, the writing of Spearman's coefficient of rank correlation can be simplified, as a result of which a more convenient formula is obtained:

$$\rho = 1 - \frac{6s_d}{n(n^2 - 1)}, \quad \text{(II.3.7)}$$

where s_d is calculated according to:

$$s_d = \sum_{i=1}^{n} (x_i - y_i)^2 = \sum_{i=1}^{n} d_i^2. \quad \text{(II.3.8)}$$

The coefficient of rank correlation takes values within the interval [–1, +1]. In analogy with the coefficient of linear correlation, if $\rho = 0$ there is no relation between the rank according to the two indicators. When the ranking of the objects fully coincides, then $\rho = +1$. In case of fully opposing ranking, $\rho = -1$.

Example

For the earlier example, the coefficient of rank correlation has to be determined as the opinion of one more expert has to be included. The ranking of the factors is given in Table II.3.2, as x_i is the ranking of the expert in machines and devices in sewing production from Table II.3.1 and y_i is of a technologist from a sewing factory.

Table II.3.2. Computation of Spearman's coefficient of rank correlation

i	1	2	3	4	5	6	7	8	9	10
x_i	7	5	2	4	1	6	8	3	9	10
y_i	4	3	5	8	1	6	7	2	10	9
d_i	3	2	–3	–4	0	0	1	1	–1	1
d_i^2	9	4	9	16	0	0	1	1	1	1
Common rank	5	4	3	7	1	6	8	2	9–10	9–10

(Continued)

(Continued)

The common rank is formed and it corresponds to the summary ranks of the two experts. In order for this ranking to be used, verification has to be made whether the opinions are close enough. A calculation is made for $s_d = 42$ and for the coefficient of rank correlation:

$$\rho = 1 - \frac{6 \times 42}{10(10^2 - 1)} = 0.75.$$

The value of the coefficient is high enough which shows that their opinion is coordinated and the common rank can be used.

Experts cannot always rank the objects in such a way that each of them is classified in a separate position. Often they give the same rank to a few objects, that is, it is possible that two or more ranks coincide at the expense of another rank that they do not award. In such case, *ranking with coinciding ranks* is present.

Example

As a part of the same research, another expert has ranked the factors as shown in Table II.3.3.

Table II.3.3. Ranking with coinciding ranks

i	1	2	3	4	5	6	7	8	9	10
x_i	5	3	2	3	1	4	5	2	6	5
x_i'	8	4.5	2.5	4.5	1	6	8	2.5	10	8

As it can be seen, two of the factors are classified in the second position, two on the third, and three on the fifth. In case of coinciding ranks, a recalculation is done for which there is a condition that the sum of the ranks in the row remains constant and equals the sum of the numbers from 1 to n:

$$\sum_{i=1}^{n} x_i' = \sum_{i=1}^{n} i = \frac{1}{2}n(n-1). \qquad \text{(II.3.9)}$$

In the last row of Table II.3.3, the recalculated ranks x_i' are given. Factors 3 and 8 have rank 2 which means that they share second and third position, respectively the recalculated ranks are:

$$x_3' = x_8' = \frac{1}{2}(2+3) = 2.5.$$

In the same way, the ranks for factors 2 and 4 are:

$$x_2' = x_4' = \frac{1}{2}(4+5) = 4.5.$$

(Continued)

The rank of factor 6 becomes $x_6' = 6$ and of factors 1, 7, and 10:

$$x_1' = x_7' = x_{10}' = \frac{1}{3}(7+8+9) = 8.$$

The rank of factor 9 becomes $x_9' = 10$.

$$\text{Verification: } \sum_{i=1}^{n} x_i' = \sum_{i=1}^{n} i = 8+4.5+2.5+4.5+1+6+8+2.5+10+8$$
$$= 1+2+3+4+5+6+7+8+9+10 = 55.$$

For calculation of the coefficient of rank correlation, Spearman's formula is modified, taking into account the presence of coinciding ranks:

$$\rho = \frac{\frac{1}{6}n(n^2-1) - s_d - U_1 - U_2}{\sqrt{\left[\frac{1}{6}n(n^2-1) - 2U_1\right]\left[\frac{1}{6}n(n^2-1) - 2U_2\right]}}, \tag{II.3.10}$$

where U_1 and U_2 are variables taking into account the presence of coinciding ranks in the first and the second row (according to the separate criteria) of the output table (before the recalculation).

$$U_j = \frac{1}{2}\sum_{i=1}^{n} u_i(u_i - 1), \tag{II.3.11}$$

where u_i is the number of repetitions of the ith rank.

Assessment of the Significance of the Rank Correlation Coefficient

For an assessment of the significance of the rank correlation coefficient ρ for $n \le 10$, the table in Appendix 7 is used. For each value of $n = 4, 5, \ldots, 10$ in the table, the possibility that the real value of s_d is lower or equal to the calculated value of s_d is given.

If the probability p, accounted from the table in Appendix 7, is lower than the given level of significance, for example $\alpha = 0.05$ or $\alpha = 0.01$, then the rank correlation coefficient is significant. If the correlation is negative, that is, $s_d > \frac{n(n^2-1)}{6} = s_0$, because of the symmetry in the distribution of s_d, the variable is determined:

$$s_d' = 2s_0 - s_d. \tag{II.3.12}$$

The assessment of significance in such cases is determined in accordance with s_d'. If $n \ge 10$, as the rule for distribution of the rank correlation coefficient (according to

Kendall and Spearman) is symmetrical, it can be approximated with a normal one. In the absence of coinciding rows, the distribution dispersion of Spearman's rank correlation coefficient is

$$\sigma_{\rho}^{2} = \frac{1}{n-1}. \quad \text{(II.3.13)}$$

Example

For the above example using Table II.3.2, a verification has to be done whether the calculated rank correlation coefficient $\rho = 0.75$ is significant.

With $n = 10$ and $s_d = 42$ from Appendix 7, the following is accounted:

$$p = \frac{0.010 + 0.007}{2} = 0.0085,$$

as 0.010 is the probability p corresponding to $s_d = 44$ and 0.007 is the probability p corresponding to $s_d = 40$. This means that with confidence level of $1 - 0.0085 = 0.9915 > 0.99$ we can accept that the rank correlation coefficient $\rho = 0.75$ is significant.

COEFFICIENT OF CONCORDANCE

In the choice of target functions and essential factors, the task for assessment of the mutual correlation relation between a few ranked rows often arises. If n objects ($i = 1, 2, \ldots, n$) are ranked according to the indicator x by m number of experts ($j = 1, 2, \ldots, m$), the data from the ranking can be presented in the form of a matrix of the ranks x_{ij} (Table II.3.4).

Table II.3.4. Matrix of the ranks

Experts *j* \ **Objects *i***	**1**	**2**	**...**	***n***
1	x_{11}	x_{21}	...	x_{n1}
2	x_{12}	x_{22}	...	x_{n2}
...	...	...	...	...
m	x_{1m}	x_{2m}	...	x_{nm}
$\sum_{j=1}^{m} x_{ij}$	$\sum_{j=1}^{m} x_{1j}$	$\sum_{j=1}^{m} x_{2j}$	...	$\sum_{j=1}^{m} x_{nj}$

The calculation of the rank correlation coefficient for each couple of experts is laborious and it also does not allow an assessment of the coordination of all experts' opinions. For an assessment of the common coordination, the *coefficient of concordance*, W, is determined. It is calculated when the summary ranks of each object are determined (last row of Table II.3.4). Then, the average value of the summary ranks is determined:

$$a = \frac{1}{n}\sum_{i=1}^{n}\left(\sum_{j=1}^{m} x_{ij}\right) = \frac{1}{2}m(n+1). \tag{II.3.14}$$

As it can be seen, it only depends on the number of the objects and the number of the experts. From the summary ranks and the variable a, s_d is determined:

$$s_d = \sum_{i=1}^{n}\left[\sum_{j=1}^{m} x_{ij} - \frac{1}{2}m(n+1)\right]^2. \tag{II.3.15}$$

The coefficient of concordance is

$$W = \frac{12 s_d}{m^2 n(n^2-1)}, \tag{II.3.16}$$

and in case of coinciding rows:

$$W = \frac{s_d}{\frac{1}{12}m^2 n(n^2-1) - m\sum_{j=1}^{m} U_j}, \tag{II.3.17}$$

where U_j assesses the number of coincidences in the ranks and is determined by the formula:

$$U_j = \frac{1}{12}\sum_{i=1}^{n} u_{ij}\left(u_{ij}^2 - 1\right), \tag{II.3.18}$$

where u_{ij} is the number of repetitions of the ith rank in the jth ranking.

The coefficient of concordance assesses the degree of coincidence of the experts' opinions regarding the ranking of the factors/objects. In this sense, it is analogous to the coefficient of multiple correlation in the regression analysis (see Section II.5). It takes values from 0 to 1 as, the higher the value of W, the more confident the researcher can be to use the results of the inquiry. In the selection of essential factors, he can rank them according to the summary ranks. During the research, he will vary the factors with the lowest ranks and will ignore the ones with the highest.

Assessment of the Significance of the Coefficient of Concordance

With fewer objects and inquired experts, the verification is done with the help of the variable:

$$F = \frac{1}{2}\ln\frac{(m-1)W}{1-W}, \tag{II.3.19}$$

which has a distribution corresponding to the one of Fisher with degrees of freedom:

$$f_1 = n - 1 - \frac{2}{m} \quad \text{and} \quad f_2 = (m-1)f_1. \tag{II.3.20}$$

With $n > 7$ the following variable is used:

$$V = m(n-1)W, \tag{II.3.21}$$

which has χ^2-distribution with degrees of freedom $f = n - 1$. When there are coinciding ranks in the ranking matrix, the variable V is calculated by the formula:

$$V = \frac{s_d}{\frac{1}{12}mn(n+1) - \frac{1}{n-1}\sum_{j=1}^{m} U_j}. \tag{II.3.22}$$

Example

The factors influencing the quality of the seam of light fabrics have to be ranked. A questionnaire has been prepared which includes 10 factors. Each factor is given a number presented in Table II.3.5.

Table II.3.5. Numbering of the factors

Number	Factor
1	Tension of the upper thread
2	Tension of the lower thread
3	Density of the seam
4	Sewing speed
5	Power of the presser foot
6	Type of the transport mechanism
7	Opening of the needle plate
8	Type of the presser foot
9	Brakes for straining the thread
10	Number of the needle

The questionnaire is filled in by 10 experts—technologists, mechanics of sewing machines in sewing companies, and teachers in sewing technologies. Each expert (marked with a letter from A to J) has given an evaluation from 1 to 10 of the degree to which he or she thinks each separate factor influences the seam quality (evaluation 1—lowest influence and evaluation 10—highest influence). The ranking of the factors by each expert is presented in Table II.3.6.

(*Continued*)

Table II.3.6. Experts' ranking of the factors

Factor	1	2	3	4	5	6	7	8	9	10
A	1	6	7	5	2	1	3	6	5	1
B	1	1	1	3	3	5	1	5	5	1
C	1	1	4	2	1	2	3	2	4	2
D	1	5	6	4	3	3	3	3	3	2
E	2	2	4	3	2	2	5	3	3	1
F	3	3	8	8	5	3	6	7	3	5
G	1	6	7	5	3	3	1	5	8	1
H	3	3	4	8	3	2	3	3	5	4
I	3	5	6	5	4	5	4	4	5	5
J	1	1	3	7	4	5	6	6	7	2

First, according to formula (II.3.14), the average value of the summary ranks is determined:

$$a = \frac{1}{10}\sum_{i=1}^{10}\left(\sum_{j=1}^{10} x_{ij}\right) = \frac{1}{2}10(10+1) = 55,$$

and using that, according to formula (II.3.15), s_d:

$$s_d = \sum_{i=1}^{10}\left[\sum_{j=1}^{10} x_{ij} - a\right]^2 = (1444+484+25+25+626+576+400+121+49+961) = 4710.$$

As we have coinciding ranks the coefficient of concordance is determined according to formula (II.3.17). According to formula (II.3.18), for each expert (j = 1, 2, …, m) the variable U_j is determined in advance. Expert number 1 has used rank 1 three times and rank 5 and rank 6 two times each. Therefore,

$$U_1 = \frac{1}{12}\left[3(3^2-1)+2(2^2-1)+2(2^2-1)\right] = 3.$$

In the same way, the variables U_2 to U_{10} are calculated.

$U_2 = 12.5$; $U_3 = 7.5$; $U_4 = 10$; $U_5 = 7$; $U_6 = 6$; $U_7 = 3$;

$U_8 = 10.5$; $U_9 = 12$; $U_{10} = 1.5$.

(*Continued*)

(Continued)

The sum of them is calculated:

$$10\sum_{j=1}^{10} U_j = 10(3+12.5+7.5+10+7+6+3+10.5+12+1.5) = 730.$$

The coefficient of concordance:

$$W = \frac{4710}{\frac{1}{12}10^2 \times 10(10^2-1) - 10\sum_{j=1}^{10} U_j} = \frac{4710}{8250-730} = 0.626.$$

Its significance is verified with a calculation by formula (II.3.22) of the auxiliary variable V:

$$V = \frac{4710}{\frac{1}{12}10 \times 10(10+1) - \frac{1}{10-1} \times 73} = 70.41.$$

The obtained value is compared with the value of the Chi-squared distribution with a level of significance $\alpha = 0.01$ and degrees of freedom $f = 10 - 1 = 9$. As $V = 70.41$ and $\chi_T^2 = 16.92$, in such case we can claim that there is a coordination of the experts' opinions. In Figure II.3.2, the ranking of the factors according to the summarized opinion of the experts is presented.

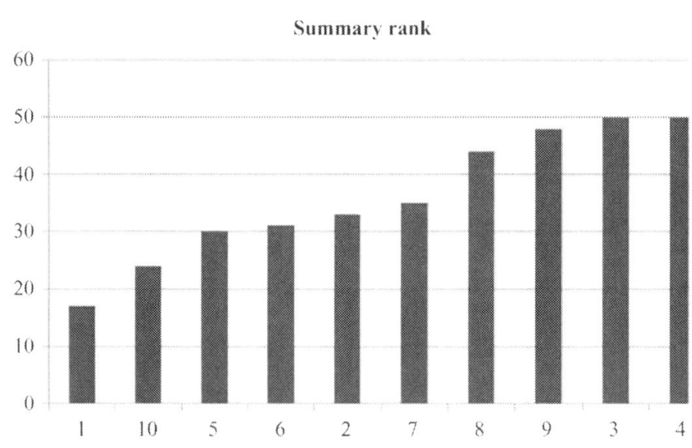

Figure II.3.2. Ranking of the factors according to the summarized opinion of the experts.

RANDOM BALANCE METHOD

Screening of the essential factors can be done with an analysis of the significance of the regression coefficients in the given model. This approach for screening of factors, however, is unsuitable as it requires the conducting of many tests and calculations. What is more appropriate is the use of plans of the experiment in which the number of the factors k is bigger than the number of observation N. These plans are used in the so-called *random balance method.* The main concept of this method is the approximate evaluation of the coefficients of the factors and their interactions on the grounds of which the essential can be separated. The design most often is made on two levels, as a result of the linear members and the interactions of second order.

The experimental plan is constructed from the full factorial experiment on the grounds of a random rule. With the random balance method, the suggestion is that the number of the essential factors l is much smaller than the number of the tested factors k, that is, the influence of the main part of the factors on the output parameter is insignificant and these factors can be changed with the equivalent disturbing influence ε.

The mathematical model is divided into two parts:

$$y = \eta + \varepsilon = \beta_0 + \beta_1 x_1 + \beta_2 x_2 + \cdots + \beta_l x_l + \xi, \tag{II.3.23}$$

where:

$$\xi = \beta_{l+1} x_{l+1} + \beta_{l+2} x_{l+2} + \cdots + \beta_k x_k + \varepsilon. \tag{II.3.24}$$

The models are constructed from the collected N experimental data. With their help and in the way described in Section II.6, the coefficients β_1, β_2, ..., β_l are determined and evaluated. Ignoring the mutual influence of the factors x_{l+1}, x_{l+2}, ..., x_k, the residual dispersion can be written as:

$$\sigma_\xi^2 \approx \beta_{l+1}^2 \sigma_{l+1}^2 + \beta_{l+2}^2 \sigma_{l+2}^2 + \cdots + \beta_k^2 \sigma_k^2 + \sigma_\varepsilon^2, \tag{II.3.25}$$

where $\sigma_{l+1}^2, \sigma_{l+2}^2, \ldots, \sigma_k^2$ are the dispersions of the respective factors.

As can be seen from formula (II.3.25), the residual dispersion σ_ξ^2 is bigger than the dispersion of the single test σ_ε^2. That is why the evaluations of the coefficients β_1, β_2, ..., β_l are with bigger mistakes than with a full factorial experiment. The advantage of the screening experiment is that it is easier to separate the dominant factors from among a big number of examined ones.

The stages of work with the random balance method are the following:

1. design of the experiment;
2. construction of a dispersion diagram; and
3. separation of essential factors.

DESIGN OF THE EXPERIMENT

The plan of the experiment is constructed in the following order. The output factors are divided into groups of two to four factors completely formally or depending of the specificity of the object. For each group, a plan of the experiment is constructed—a full or fractional factorial experiment. The general plan of the experiment is constructed by a random (e.g., by tables of random numbers) mixing of the rows of the separate group plans.

Example

The essential factors of four main factors (x_1, x_2, x_3, x_4) and six interaction of second order (x_1x_2, x_1x_3, x_1x_4, x_2x_3, x_2x_4, x_3x_4) have to be separated on the ground of six tests.

The factors are divided into two groups: first group, x_1 and x_2, and second group, x_3 and x_4. For each group, matrices are constructed for a full factorial experiment of the type 2^2 (Table II.3.7). On the basis of the two matrices, a common matrix of the experiment is constructed for $N = 6$ as the rows of the two matrices are selected in a random way. Rows 3, 1, 4, 1, 2, and 4 are taken from the first matrix and rows 1, 4, 2, 3, 2, and 1 from the second one.

Table II.3.7. Full factorial experiment for both groups

Number of row \ Factor	x_1	x_2	Number of row \ Factor	x_3	x_4
1	+	+	1	+	+
2	+	–	2	+	–
3	–	+	3	–	+
4	–	–	4	–	–

The common matrix of the experiment is given in Table II.3.8. The values of the interaction of the second order are also given in it and in the column before the last the values of the output parameter y obtained by the experiment are given.

Table II.3.8. Common matrix of the experiment

Test number \ Factor	x_1	x_2	x_3	x_4	x_1x_2	x_1x_3	x_1x_4	x_2x_3	x_2x_4	x_3x_4	y	y'
1	–	+	+	+	–	–	–	+	+	+	31	8
2	+	+	–	–	+	–	–	–	–	+	34	12
3	–	–	+	–	+	–	+	–	+	–	5	5
4	+	+	–	+	+	–	+	–	+	–	26	4
5	+	–	+	–	–	+	–	–	+	–	7	7
6	–	–	+	+	+	–	–	–	–	+	14	14

CONSTRUCTION OF A DIAGRAM OF DISPERSION

The construction is done in the following way: the factors are entered conditionally on the abscissa axis and the values of the output parameter in the form of points are entered on the ordinate axis. These points are entered for each factor separately as they are divided into two subgroups: on the left for the tests in which the respective factor has been on the lower level (–) and on the right in the cases when the factor has been on the upper level (+). The medians are entered for each subgroup of points and for each factor (in case of even number of points—in the middle between the two inner points, in case of odd number—through the middle point). The essential factors are determined by the absolute value of the margin between the left and the right median. The bigger it is, the more essential the respective factor is.

The diagram of dispersion for the ten factors in Table II.3.8 is given in Figure II.3.3.

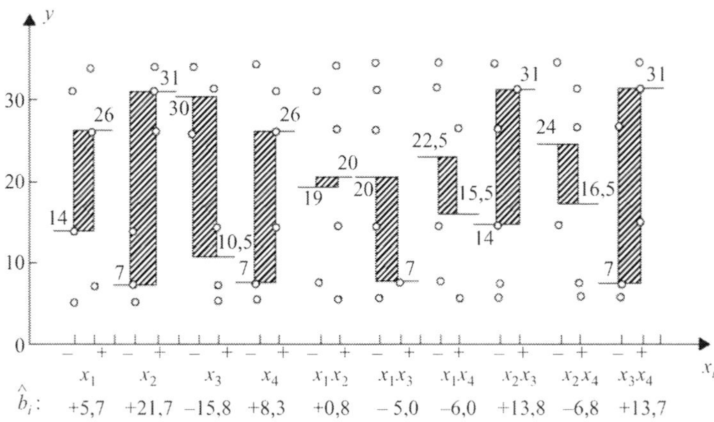

Figure II.3.3. Diagram of dispersion.

The results obtained are characteristic for the diagrams of dispersion with the random balance method. It is seen that for a big part of the factors and the interactions, the margins between the medians are essential. The margins are bigger for: x_1 (+24), x_1x_4 (+24), x_3 (–19,5) and x_4 (+19). The sign of the margin between the upper (right) median and the lower (left) median shows the direction in which the given factor influences the output parameter. If the margin is positive, the output parameter grows with the increase of the factor and vice versa.

The margins between the medians, also called *contribution of the factors*, can serve for tentative evaluation of the significance of the separate factors. However, they are not efficient for evaluation of the level of influence of the separate factors on the output parameter. The following evaluations of the coefficient are more useful:

$$\hat{b}_i = \frac{1}{N_1}\sum_{j=1}^{N_1} y_i - \frac{1}{N - N_1}\sum_{j=N-N_1}^{N} y_j, \tag{II.3.26}$$

where indices from 1 to N_1 indicate the tests in which the factor x_i is in the upper (+) level while indices from N_1 to N indicate the tests in which the factors x_i is in the lower (–) level. The change in the formula is necessary because of the fact that the number of tests N_1 in which $x_i = +1$ in most of the cases is not equal to the number of tests $(N - N_1)$ in which $x_i = -1$. For the examined example (Table II.3.8), the evaluations $\hat{b}_i$ are entered in Figure II.3.3 below the respective factors. They confirm most of the results obtained during the analysis of the contributions, for example, factor x_2 is dominant according to both approaches but the interaction x_3x_4 is given a more rearward position. In addition to the contributions and the evaluations, $\hat{b}_i$, for the degree of influence of the separate factors we can also judge by the number of the so-called *separating points*.

Separating points are those points for which when $x_i = +1$, the values of the output parameter y are above the highest value of y when $x_i = -1$ and those for which when $x_i = -1$ the values of y are lower than its lowest value when $x_i = +1$. It is obvious that the higher the number of separating points is, the stronger the influence is of the respective factor on y. For example, it can be seen (Figure II.3.3) that all six points of the factor x_2 are separating. The number of separating points for the factors x_3 (four points), x_3x_4 (four points), and x_4 (three points) is also significant.

There also is an approach in which the interactions of second order can be selected without constructing a diagram of dispersion for them. This is especially important in case of a great number of factors because then the number of interactions grows very fast, followed by the quantity of calculation operations and graphical constructions. The approach is based on the fact that the essential interaction x_ix_j has a big number of separating points, both, on levels $+x_ix_j$ and $-x_ix_j$. In the first case, the factors x_i and x_j have to be with the same sign and in the second case with different signs. Therefore, the interaction of these factors that have separating points, both in case of the same and in case of different levels, can be examined. The last condition is implemented when the upper or the lower parts of the diagram of dispersion of the factors x_i and x_j are mirror images. For example, in Figure II.3.4 the lower parts of the diagrams of dispersion of x_i and x_j are mirror images of one another.

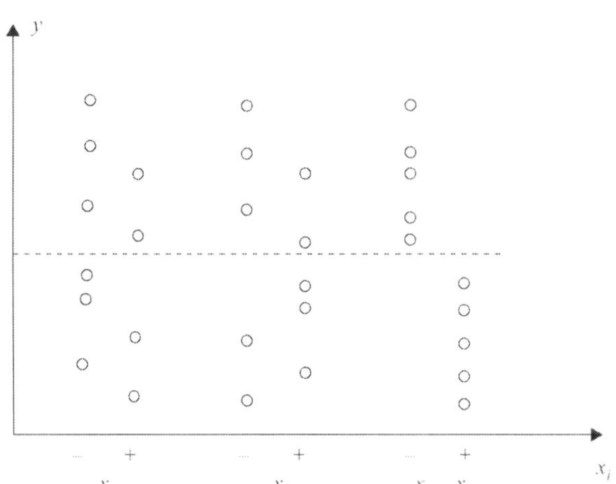

Figure II.3.4. Diagrams of dispersion of x_i and x_j.

For the interaction x_ix_j, these points turn out to be on a lower level. The points recur in the upper part of the two diagrams. For x_ix_j they turn out to be on an upper level. As a result of the interaction x_ix_j has separating points on both levels. This figure graphically shows how essential interaction can be found when the factors themselves turn out to be nonessential (both factors have two separating points only). The picture is analogous in Figure II.3.3 for factors x_3 and x_4. The two uppermost points are repeated and the two lowermost points are symmetrical. The interaction x_3x_4 has respectively two separating points, both in the upper and in the lower parts of the diagram.

SEPARATION OF ESSENTIAL FACTORS

Separation of the essential factors on the diagrams of dispersion has to be done very carefully as the number of tests is small and the design matrix is not strictly orthogonal. That is why the separation of the factors in done consecutively, one by one or a few at a time.

After the separation of the chosen factors, their influence is eliminated as they are fixed at one level. If the separated factor $x_i = +1$, the variable $\hat{b}_i$ is deducted from the values of y. Thus the factor x_i is conditionally transferred on to level $x_i = -1$ for the whole experiment. If the separated factor $x_i = -1$, the variable $\hat{b}_i$ is added. Thus the factor x_i is transferred on to level $x_i = +1$. When a few factors are separated simultaneously, a simultaneous transfer is implemented for all separated factors.

After a correction of the values of the output parameter, the diagrams of dispersion are constructed again for the other factors. Another group of factors is separated and the values of y are corrected again. This consecutive separation goes on until the contribution of the rest of the factors, or the evaluations $\hat{b}_i$, turn out to be close. In cases when the contributions of a few factors on the diagrams of dispersion are close, the factors having evaluations $\hat{b}_i$ higher than the absolute value for which the number of separating point is bigger have to be chosen. It should not be forgotten though that in accordance with its main precondition, the random balance method allows for only a small number of essential factors to be separated.

Regarding the example: in accordance with the above given recommendations, x_2 is separated as an essential factor as with similar contributions with x_3x_4 and a number of separating points, $\hat{b}_2 > \hat{b}_3$. The factors x_2 is transferred on to a lower level as $x_2 = +1$. From the values of y in test numbers 1, 2, and 4 we deduct the value of the evaluation $\hat{b}_2 = +21.7 \approx +22$. The values of y obtained after the correction are given in the last column of Table II.3.8 and are marked with y'. With these values, the diagrams of dispersion are constructed again for the remaining nine factors (Figure II.3.5).

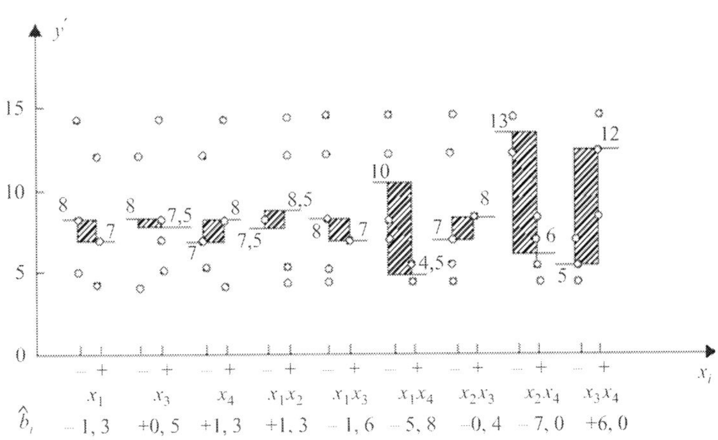

Figure II.3.5. Reconstructed diagram of dispersion for the remaining factors.

The contributions and the values of the evaluations $\hat{b}_i$ are given in the figure. The most significant interactions appear to be x_2x_4 and x_3x_4, as the interaction x_3x_4 was dominant in the first screening, too. The two interactions can be separated at the same time and the screening is done in the same way.

II.4. Main Stages of Experimental Modeling

Experimental modeling goes through the following main stages:

1. **Choice of target function**
   Definition of the target is determined by the tasks assigned to the given production. It cannot be done with the help of formal mathematical methods. Considerations on the choice of target functions are described in Section II.2.
2. **Choice of input factors**
   A decision is made on which of the factors will be modified during the experiment. The rest of them are fixed to constant values or their influence refers to the action of random disturbances. Requirements to the factors are described in Section II.3.
3. **Choice of the structure of the model**
   The type and the structure of the mathematical model are chosen on the basis of preliminary examination of the model. Polynomial models are used most often.
4. **Planning of the experiment**
   Data from the observations can be collected with the help of an unplanned or planned-in-advance experiment. Specific plans of the experiment are described in Sections II.6, II.7, and II.11.
5. **Realization of the tests**
   Technological and technical realization of the task. Methods for its solving are determined by the specific characteristics of the object.
6. **Processing of experiment data**
   The coefficients participating in the mathematical model are determined on the grounds of the data obtained by the experiment.
7. **Statistical analysis of the model**
   Coincidence of the values predicted according to the model and the ones determined by the test is checked. If the coincidence is good, it is said that the model is adequate and can be used for optimization. Otherwise, the structure of the model is changed or new factors are included in the investigation. A new experiment can be executed or the data for obtaining the new model structure is processed during the transition to a new model. At the stage of statistical analysis, conclusions are made for the level of influence of the factors on the output variable.

8. **Optimization**
   The best technological regime, in a predefined sense, is determined with the help of the model.
9. **Verification of the optimal regime and technological interpretation of the research results**
   The determined optimal regime is realized and the coincidence of the obtained results with the results predicted by the model is checked. Results are rationalized from technological point of view. Graphs are drawn showing the variation of the output variable depending on the factors.

As it has been noted, statistical models are formed by using results of an experiment, that is, a measurement is realized. Due to the presence of *errors from the measurement*, the real value of the measured variable x cannot be determined. Only the interval in which it is situated is determined:

$$x = x^* \pm u, \tag{II.4.1}$$

where x^* is the measured value;
u is total error (the sum of all errors).

Errors may be:

- *Errors from the choice:* They occur when not the whole population was examined but just a part of it, the so-called sample. The values obtained differ from the real ones but the errors can be defined with the help of mathematical statistics. They are evaluated with the absolute q and the relative p confidence error, the determination of which is clarified in Section I.3.
- *Errors from the measurements:* They are obtained as a result of an imperfection in the construction of the device or in its execution. The permissible error is specified in the passport and is accounted in registration of the results.
- *Random errors:* They occur under the influence of diverse factors and cannot be avoided. However, with the help of the probability theory and mathematical statistics, the investigator can recognize and reduce their influence to a significant extent. The approach of randomization is used often, in which tests are conducted in a random (not arranged) way with the aim of eliminating factors such as variations in temperature, humidity, and efficiency of the workers.
- *Systematic errors:* They occur as a result of permanently acting reasons, for example, a device out of order or incorrect methods for examination. They have the same value in all tests and are hard to find. Special tests are necessary for them to be found and eliminated, for example, measurement of the same variable with a few measurement instruments or measurement with the same instrument of a few standards. Systematic errors can be eliminated with the introduction of corrections in the observation results, for example, the Pressley's correction coefficient in determination of the strength of cotton fibers.
- *Gross errors:* They occur in case of an incorrect account of the scale, registration, or calculation. They usually are the easiest to find as they differ very much from the rest of

the results. The obtained result should be eliminated and the test should be repeated, if possible.

A criterion for the presence of a gross error is when the value of the examined variable is out of the confidence interval $(\overline{x} \pm t \cdot \sigma)$—in a known dispersion of the population, and respectively, $\left(\overline{x} \pm t \cdot \frac{S}{\sqrt{n}}\right)$—in case of a work with a sample with volume n.

In statistical modeling it is accepted that systematic and gross errors are eliminated.

II.5. Regression Analysis

Regression analysis is one of the first methods for construction and evaluation of models of multifactorial objects based on the results of experiments. *Regression* or *regression equation* is the relation between the random variable, Y, and one or more variables $X_1, X_2, ..., X_m$, which may be random or nonrandom. Regression is the conditional mathematical expectation of a random variable Y, provided that $X_1, X_2, ..., X_m$ have certain predefined values $x_1, x_2, ..., x_m$, i.e., $M(Y / X_1 = x_1, X_2 = x_2, ..., X_m = x_m)$.

If Y is a continuous differentiable function of the factors, it can be situated in a Taylor series and, respectively, be approximated with polynomials. Polynomial equations are appropriate when there are no abrupt variations of the output variable in a change of the input factors. This requirement is met in most of the cases in textile production. The general form of the polynomial regression model is

$$y = \beta_0 + \sum_{i=1}^{m} \beta_i x_i + \sum_{i=1}^{m-1} \sum_{j=i+1}^{m} \beta_{ij} x_i x_j + \sum_{i=1}^{m} \beta_{ii} x_i^2 + \cdots \tag{II.5.1}$$

Polynomial models can be used regardless of whether the factors are coded, normalized, or in natural units. They are the most widely accepted but are not the only regression models. Regression can be presented with random functions, f_i, of the factors $x_1, x_2, ..., x_m$:

$$y = \sum_{i=1}^{k} \beta_i f_i. \tag{II.5.2}$$

The first task of regression analysis after the determination of the type of functions, f_i, is finding the *coefficients of the regression equation.* They are determined from the experimentally obtained results. It is convenient to arrange the data as shown in Table II.5.1.

The real value of y is not known. What the experimenter has is the measured value of y at the uth measurement y_u. For this reason, the real values of the parameters β_i cannot be found, but just their evaluations b_i. If in formula (II.5.2) the coefficients β_i are replaced with the evaluations b_i, the following equation would be obtained:

$$\hat{y}_u = \sum_{i=1}^{k} b_i f_{iu}. \tag{II.5.3}$$

Table II.5.1. Data arrangement for conduction of regression analysis

Number	x_1	x_2	x_3	...	x_m	y
1	x_{11}	x_{21}	x_{31}	...	x_{m1}	y_1
2	x_{12}	x_{22}	x_{32}	...	x_{m2}	y_2
...	...	...	...	...	...	...
u	x_{1u}	x_{2u}	x_{3u}	...	x_{mu}	y_u
...	...	...	...	...	...	...
N	x_{1N}	x_{2N}	x_{3N}	...	x_{mN}	y_N

It is obvious that the evaluations of the regression coefficients will be as much more exact as the margins ($y_u - \hat{y}_u$) called residuals are smaller. This is ensured by the requirement for the sum Q_{res} to be minimal. Q_{res} is the residual sum of the squares and shows the influence of inadequacy and random error. It is defined as:

$$Q_{res} = \sum_{u=1}^{N} \left(y_u - \hat{y}_u \right)^2, \tag{II.5.4}$$

where y_u is the measured value,

$\hat{y}_u$ is value calculated by the regression equation.

The method used is called "method of least squares."

Replacing $\hat{y}_u$ from formula (II.5.3) in formula (II.5.4), for the sum of the squares of the residuals we get

$$Q_{res} = \sum_{u=1}^{N} \left(y_u - b_1 f_{1u} - b_2 f_{2u} - \cdots - b_k f_{ku} \right)^2. \tag{II.5.5}$$

In order for Q_{res} to be minimized, it is defined in relation to the unknown evaluations and its first derivatives are equaled to zero:

$$\begin{aligned} \frac{\partial Q_{rez}}{\partial b_1} &= -2\sum_{u=1}^{N} \left(y_u - b_1 f_{1u} - b_2 f_{2u} - \cdots - b_k f_{ku} \right) f_{1u} = 0 \\ \frac{\partial Q_{rez}}{\partial b_2} &= -2\sum_{u=1}^{N} \left(y_u - b_1 f_{1u} - b_2 f_{2u} - \cdots - b_k f_{ku} \right) f_{2u} = 0 \\ &\cdots\cdots\cdots\cdots\cdots\cdots\cdots\cdots\cdots\cdots \\ \frac{\partial Q_{rez}}{\partial b_k} &= -2\sum_{u=1}^{N} \left(y_u - b_1 f_{1u} - b_2 f_{2u} - \cdots - b_k f_{ku} \right) f_{ku} = 0 \end{aligned} \tag{II.5.6}$$

After opening the brackets and placing of

$$
\begin{aligned}
g_{ij} &= g_{ji} = \sum_{u=1}^{N} f_{iu} f_{ju} \\
g_{ii} &= \sum_{u=1}^{N} f_{iu}^2 \\
z_i &= \sum_{u=1}^{N} f_{iu} y_u
\end{aligned}
\tag{II.5.7}
$$

the system of equations (II.5.6) looks in the following way:

$$
\begin{aligned}
& b_1 g_{11} + b_2 g_{12} + \cdots + b_k g_{1k} = z_1 \\
& b_1 g_{21} + b_2 g_{22} + \cdots + b_k g_{2k} = z_2 . \\
& \cdots\cdots\cdots\cdots\cdots\cdots\cdots\cdots\cdots\cdots \\
& b_1 g_{k1} + b_2 g_{k2} + \cdots + b_k g_{kk} = z_k
\end{aligned}
\tag{II.5.8}
$$

It is called a *system of normal equations*. When it is solved, the values of the evaluations of the regression coefficients $b_1, b_2, \ldots, b_k$ are obtained. Very often the system of normal equations is entered in a matrix form as the following notations are used:

- Matrix of the design of the experiment ($N \times m$)—every row of the matrix contains the values of the factors which are given during the tests:

$$
X = \begin{Vmatrix} x_{11} & x_{21} & \cdots & x_{m1} \\ x_{12} & x_{22} & \cdots & x_{m2} \\ \cdots & \cdots & \cdots & \cdots \\ x_{1N} & x_{2N} & \cdots & x_{mN} \end{Vmatrix}.
$$

- Extended matrix of the design ($N \times k$)—elements of each row are obtained as in the functions, $f_1, f_2, \ldots, f_k$, the values of $x_1, x_2, \ldots, x_m$ are replaced from the corresponding row of X:

$$
F = \begin{Vmatrix} f_{11} & f_{21} & \cdots & f_{k1} \\ f_{12} & f_{22} & \cdots & f_{k2} \\ \cdots & \cdots & \cdots & \cdots \\ f_{1N} & f_{2N} & \cdots & f_{kN} \end{Vmatrix}.
$$

- N-dimensional vector of the results from the tests:

$$
y = \begin{Vmatrix} y_1 \\ y_2 \\ \cdots \\ y_N \end{Vmatrix}.
$$

- Information matrix of the design ($k \times k$)—the elements of the matrix are calculated from formulae (II.5.7) and (II.5.8):

$$G = \begin{Vmatrix} g_{11} & g_{21} & \cdots & g_{k1} \\ g_{12} & g_{22} & \cdots & g_{k2} \\ \cdots & \cdots & \cdots & \cdots \\ g_{1k} & g_{2k} & \cdots & g_{kk} \end{Vmatrix}.$$

- Covariance matrix ($k \times k$)—reciprocal to the G-matrix:

$$C = G^{-1} = \begin{Vmatrix} c_{11} & c_{21} & \cdots & c_{k1} \\ c_{12} & c_{22} & \cdots & c_{k2} \\ \cdots & \cdots & \cdots & \cdots \\ c_{1k} & c_{2k} & \cdots & c_{kk} \end{Vmatrix}.$$

- k-dimensional vector of the evaluation of the regression coefficients:

$$b = \begin{Vmatrix} b_1 \\ b_2 \\ \cdots \\ b_k \end{Vmatrix}.$$

- k-dimensional vector $z = G \cdot b$:

$$z = \begin{Vmatrix} z_1 \\ z_2 \\ \cdots \\ z_k \end{Vmatrix}.$$

The evaluations obtained by the least squares method are:

- *undisplaced*—$\mathrm{M}\{b_i\} = \beta_i$, $i = 1, 2, \ldots, k$
- *consistent*—with $\mathrm{N} \to \infty$ the evaluations b_i with a probability 1 tend to the real values of the coefficients β_i
- *efficient*—the determinant of their covariance matrix is smaller than the determinant of the covariance matrix of any other evaluations, that is, $\left|\mathrm{cov}\{b\}\right| \to \min$.

When the regression equation has the following form:

$$y = b_0 + b_1 x, \tag{II.5.9}$$

that is, the model to be found is a polynomial of the first order with one factor, the evaluations of its coefficient are determined as:

$$b_1 = r_{xy}\frac{S_y}{S_x} \tag{II.5.10}$$

$$b_0 = \overline{y} - b_1\overline{x}, \tag{II.5.11}$$

where r_{xy} is the linear correlation coefficient of the normally distributed variables x and y,
$\overline{x}$ and $\overline{y}$ are the arithmetic mean values of the two random variables determined by the experiment,
S_x and S_y are the standard deviations of the variables determined by the experiment.

The statistical evaluations calculation is clarified in Section I.3. After calculation of the evaluations b_i, the model can be written but not used. Before it is applied, it should be verified whether it is adequate, that is, whether the forecasted values correspond to the measured ones to a sufficient degree. For verification of the adequacy, the residual sum of the squares Q_{res} is used, from which the residual dispersion is calculated:

$$S_{res}^2 = \frac{1}{N-k}\sum_{u=1}^{N}(y_u - \hat{y}_u)^2. \tag{II.5.12}$$

S_{res}^2 characterizes the dispersion of the experimentally determined values, y_u, around the ones calculated by the model, $\hat{y}_u$, as the degrees of freedom are obtained when the number of linear relations between the tests, k (the number of coefficients), is deducted from the number of tests N. It is compared with the evaluation of the random error S_ε^2 variance determined on the grounds of l additional tests (usually $l = 5$–10) conducted in one and the same technological regime. The verification is done according to the Fisher criterion:

$$F_R = \frac{S_{res}^2}{S_\varepsilon^2}, \tag{II.5.13}$$

as the higher of the two evaluations is written in the numerator. Usually $S_{res}^2 > S_\varepsilon^2$ because $\sigma_{res}^2 > \sigma_\varepsilon^2$. The table value of the Fisher criterion under the following conditions $F_T(a, f_1 = N-k, f_2 = l-1)$ is taken from Appendix 3 and a comparison is made.

- If $F_R \le F_T$, the model is adequate.
- If $F_R > F_T$, the model is inadequate.
- If $S_{res}^2 < S_\varepsilon^2$, the variance relation is written down:

$$F_R = \frac{S_\varepsilon^2}{S_{res}^2}. \tag{II.5.14}$$

The model is adequate if $F_R \le F_T$. If $F_R > F_T$, most likely a mistake has been made because always $\sigma^2_{res} > \sigma^2_\varepsilon$ (residual variance is caused both from inadequacy or a random disturbance). In some cases (e.g., in the event of a passive experiment), no additional test can be conducted for determination of S^2_ε. In this case, the *coefficient of multiple correlation*, *R*, can be used for evaluation of the model. It can be obtained through examination of the dispersion of the output variable around its mean value and it is determined from the total of the squares:

$$Q = Q_R + Q_{res} = \sum_{u=1}^{n} \left(y_u - \overline{y}\right)^2, \tag{II.5.15}$$

where Q_R is a sum showing the deviations from the mean value, $\overline{y}$, caused by the influence of the factors included in the regression model. It is determined by the formula

$$Q_R = \sum_{u=1}^{n} \left(\hat{y}_u - \overline{y}\right)^2. \tag{II.5.16}$$

The mean value $\overline{y}$ is determined from the measured values:

$$\overline{y} = \frac{1}{N}\sum_{u=1}^{N} y_u. \tag{II.5.17}$$

Coefficient of multiple correlation (*of determination*) is the following variable:

$$R = \sqrt{\frac{Q_R}{Q}} = \sqrt{1 - \frac{Q_{res}}{Q}}. \tag{II.5.18}$$

If the influence of all factors affecting the output parameter is taken into account, $Q = Q_R$, $Q_{res} = 0$, then the coefficient of multiple correlation $R = 1$. If the regression model does not show the influence of any of the factors correctly, then $R = 0$. The closer the value of the coefficient of multiple correlation is to zero, the greater the possibility for the model to be accurate is.

However, not only the value of *R* is an indication for the accuracy of the model. It can be 1 even when the number of tests, *N*, is equal to the number of coefficients in the model, *k*. In this case, the degrees of freedom $f_{res} = N - k = 0$ and there is no ground for verification of adequacy. The reliability of the conclusions regarding the model is higher when the number of degrees of freedom, f_{res}, is higher. For elimination of this drawback of the multiple correlation coefficient, calculation of the so-called *adjusted coefficient of multiple correlation* is suggested.

The adjusted coefficient of multiple correlation is an index showing the number of regression parameters in a function of the number of tests:

$$R_{ad} = \sqrt{1 - (1 - R^2)\frac{N-1}{N-k}}, \tag{II.5.19}$$

where *N* is the number of tests,

k is the number of regression coefficients in the model.

The regression analysis also includes the calculation of the *standard residual error*:

$$S = \sqrt{\frac{Q_{res}}{N-k}}. \qquad \text{(II.5.20)}$$

The Fisher criterion is used for evaluation of the significance of multiple correlation coefficient and its value is

$$F_R = \frac{Q_R/(k-1)}{Q_{res}/(N-k)} = \frac{R^2(N-k)}{(1-R^2)(k-1)}. \qquad \text{(II.5.21)}$$

If $F_R > F_T(\alpha, f_1 = f_R = k-1, f_2 = f_{res} = N-k)$, the correlation coefficient is significant (the model is adequate) and vice versa—if $F_R \le F_T$, the correlation coefficient is insignificant (the model is inadequate). Sometimes the absolute value of some of the evaluations of the coefficients in the regression model is close to zero and this, most likely, is due only to the variations caused by the random character of phenomena. Such evaluations are insignificant and can be excluded from the model, thus increasing the degrees of freedom, f_2.

The verification of the coefficients of regression equation's significance is brought to verification of the equality of a given value to zero and is done with the Student's *t*-criterion. The following variable is calculated for all independent coefficients:

$$t_i = \frac{b_i}{\sqrt{c_{ii} \cdot S_\varepsilon^2}}, \qquad \text{(II.5.22)}$$

where c_{ii} are the values of the diagonal elements from the covariance matrix,
S_ε^2—the random error variance determined by l additional tests.

The coefficients of the regression equation are insignificant when:

$$|t_i| \le t\left(\frac{\alpha}{2}, f = l-1\right). \qquad \text{(II.5.23)}$$

The verification is not correct for those coefficients of the regression equations, b_i, which have nonzero covariance with other evaluations, that is, some of the elements of the covariance matrix, $c_{ij} \ne 0$. In the full and the fractional factorial experiments (Sections II.6 and II.7), the matrix is diagonal and the diagonal elements are equal to each other:

$$c_{ii} = \frac{1}{N}, \qquad \text{(II.5.24)}$$

where N is the number of tests in the design.

Example

The irregularity of a cotton sliver is measured at the exit of drawing frame as the work has been done in 12 different technological regimes. The movement of the fibers in the drawing device is simulated with a computer program and the forecasted values of irregularity are calculated. The dependency between the measured CV_{meas} and the forecasted (from the simulation) results shown in Table II.5.2 should be determined.

Table II.5.2. Measuring and simulation results

CV_{meas}	5.91	6.18	6.00	5.94	6.17	5.85	6.49	6.70	6.18	6.35	6.16	5.95
CV_{sim}	5.61	5.93	5.67	5.70	5.80	5.52	6.32	6.29	5.93	5.90	5.83	5.70

The problem is broken down to the determination of a relation between one input and one output parameter, that is, to the regression equation from the type (II.5.9). The measured irregularity is chosen for an input parameter x, and the prognosis for an output parameter y. Its evaluations and coefficients are determined according to formulae (II.5.10) and (II.5.11).

The following are determined in advance:

The mean value of x: $\overline{x} = 6.157\%$
The mean value of y: $\overline{y} = 5.850\%$
The standard deviation of x: $S_x = 0.255\%$
The standard deviation of y: $S_y = 0.248\%$
The linear correlation coefficient of x and y (according to formula (I.5.1)): $r_{xy} = 0.95$.

Afterward, the coefficients are determined:

$$b_1 = r_{xy}\frac{S_y}{S_x} = 0.95\frac{0.248}{0.255} = 0.925 \quad \text{and}$$

$$b_0 = \overline{y} - b_1\overline{x} = 5.85 - 0.925 \times 6.157 = 0.155.$$

The dependency between the two random variables and the regression equation are shown graphically in Figure II.5.1.

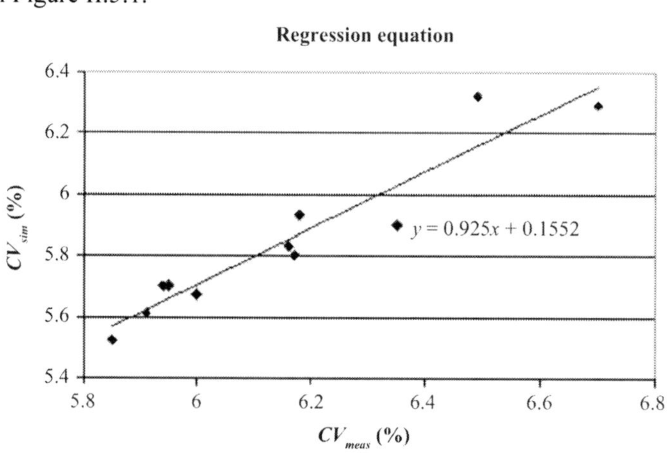

Figure II.5.1. Graphical dependency between the measuring and simulation results.

Example

The influence of the following characteristics of fibers on the specific yarn tenacity, y (cN/tex), should be determined:

x_1—staple length, mm,
x_2—Pressley index of strength,
x_3—micronaire value,
x_4—trash quantity, %,
x_5—short fiber content, %.

The data for the fibers and the mean values obtained for the yarn tenacity are shown in Table II.5.3.

Table II.5.3. Measured fiber and yarn characteristics

Variant	Staple length	Pressley index	Micronaire	Trash	Short fibre content	Yarn tenacity
1	33.19	8.46	3.5	2.49	12.56	9.7
2	32.97	8.12	3.4	2.79	13.75	10.1
3	33.11	7.38	3.5	3.11	11.37	10.1
4	33.18	8.03	3.5	3.18	12.25	9.7
5	33.08	8.07	3.6	3.37	11.75	9.7
6	33.65	7.69	3.6	3.40	12.35	10.4
7	33.63	8.11	3.6	3.58	12.51	10.4
8	32.65	8.26	3.4	4.35	15.02	9.5
9	32.70	8.05	3.5	3.41	15.55	10.2
10	33.66	7.73	3.4	3.30	13.98	10.8

A universal program as Microsoft Office Excel can be used for derivation of a linear regression model. The subprogram making the necessary calculations and presenting them in the form of tables is called Regression and is in the Data Analysis menu.

After determination of the location of input data, the desired confidence level (95%), and the initial cell for situation of the output tables, the program derives the following results given in Table II.5.4.

Table II.5.4. Summary output from Excel

Summary Output	
Regression statistics	
Multiple R	0.976354609
R Square	0.953268322
Adjusted R Square	0.894853724
Standard Error	0.132556067
Observations	10

(Continued)

(*Continued*)

ANOVA

	df	SS	MS	F	Significance F
Regression	5	1.433715556	0.28674311	16.3190086	0.009111789
Residual	4	0.070284444	0.01757111		
Total	9	1.504			

	Coefficients	Standard error	*t*-Stat	*P*-value	Lower 95%	Upper 95%
Intercept	−18.6137336	5.489590201	−3.3907328	0.02751139	−33.8553111	−3.37216
X Variable 1	0.843816236	0.143665275	5.87348776	0.00419737	0.444936659	1.242696
X Variable 2	−0.88445364	0.193219969	−4.5774443	0.01020315	−1.42091939	−0.34799
X Variable 3	1.41853217	0.795481845	1.78323638	0.14912085	−0.79008408	3.627148
X Variable 4	−0.27125565	0.108129	−2.5086299	0.06615402	−0.5714705	0.028959
X Variable 5	0.281201747	0.05369104	5.23740545	0.00635149	0.132131212	0.430272

In the upper part, the results of the regression analysis are shown: the coefficient of multiple correlation, the square of the multiple correlation, the adjusted coefficient of the multiple correlation, the standard error, and the number of observations on grounds of which the calculations are done. In the middle part, the results from the dispersion analysis (ANOVA) are shown: the degrees of freedom, the sums of the squares (from the regression, the residual, and the total), the dispersions (from the regression and the residual), the accounted, and the table values of the Fisher distribution. In the lower part of the table, the calculated values of the regression coefficients (the 0 and 1 and the ones before the separate factors) are shown as well as the standard errors, the values t_r, the percentage error, the upper and the lower bounds of the regression coefficients' evaluations with a confidence level of 95%.

As the coefficient of the multiple correlation is very high, $R = 0.976$, and $F_R = 16.319 > F_T = 0.009$, the model obtained is adequate and can be written in the following way:

$$y = -18.614 + 0.844 \times x_1 - 0.884 \times x_2 + 1.419 \times x_3 - 0.271 \times x_4 + 0.281 \times x_5.$$

It can be seen from it that the strongest influence on the yarn tenacity is the fineness of the fibers (the micronaire value) which determines the number of fibers in the section of the yarn, followed by the strength of the fibers (the Pressley index), and their staple length, and the weakest influence—the content of impurities and short fibers.

As the sign of the zero coefficient is negative, the direction of influence of the separate factors is opposite to the sign before the coefficient, that is, the tenacity increases with an increase of the strength of the fibers, decrease in the micronaire value, and so on.

II.6. Full Factorial Experiment

Often, for description of experimental conditions models are used in which factors enter only in the first degree or in products with other factors (in interactions):

$$\hat{y} = b + \sum_{i=}^{n} b_i x_i + \sum_{i>j} b_{ij} x_i x_j + \sum_{i>j>k} b_{ijk} x_i x_j x_k + \cdots \tag{II.6.1}$$

For derivation of the model, it should be possible for all factors to be modified independently of each other and they should be coded in advance by formula (II.3.3). When coded in this way, the area of variance of the factors is determined by the inequalities:

$$-1 \leq x_i \leq 1, \quad i = 1, 2, \ldots, m. \tag{II.6.2}$$

Full factorial experiment (FFE) is an experiment in which all possible combinations on two levels of the factors (upper +1 and lower −1) are realized. The number of these combinations for m factors is

$$N = 2^m. \tag{II.6.3}$$

The design of the experiment with two input factors is given in Table II.6.1 and with three factors in Table II.6.2.

The following rule is convenient to be applied for its preparation: the signs of the factor x_1 change in an interval of one level, of x_2 in an interval of two levels, of x_3 in an interval of four levels and so on, on the degrees of the number 2. The sign (+) corresponds to +1 and the sign (−) corresponds to −1. The column for x_0 always equals 1 as x_0 is a fictitious variable introduced because of b_0. The examined designs have a simple geometrical interpretation (Figure II.6.1). The area of variance of the factors with (a) two input factors is a square and (b) with three factors a cube. The points of the design correspond to the vertices of the square/cube.

Table II.6.1. Design of the experiment with two input factors

Number	x_0	x_1	x_2	y
1	+	–	–	y_1
2	+	+	–	y_2
3	+	–	+	y_3
4	+	+	+	y_4

Table II.6.2. Design of the experiment with three input factors

Number	x_0	x_1	x_2	x_3	y
1	+	–	–	–	y_1
2	+	+	–	–	y_2
3	+	–	+	–	y_3
4	+	+	+	–	y_4
5	+	–	–	+	y_5
6	+	+	–	+	y_6
7	+	–	+	+	y_7
8	+	+	+	+	y_8

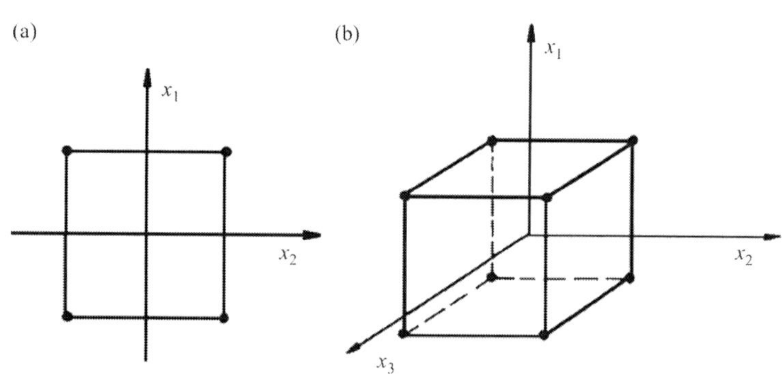

Figure II.6.1. Geometrical interpretation of experimental points.

PROPERTIES OF THE EXTENDED MATRIX OF FFE

1. The total of the elements of each column (factor) equals zero:

$$\sum_{j=1}^{N} x_{ij} = 0 \qquad \text{for } i = 1, 2, 3, \ldots, m\text{—number of the factor.} \tag{II.6.4}$$

2. The total of the squares of the elements of each column equals the number of tests N:

$$\sum_{j=1}^{N} x_{ij}^2 = N \qquad \text{for } i = 0, 1, 2, \ldots, m. \tag{II.6.5}$$

3. The total of the element products of any two columns equals zero:

$$\sum_{j=1}^{N} x_{ij} x_{lj} = 0 \qquad \text{for } i, l = 0, 1, 2, \ldots, m \ (i \neq l). \tag{II.6.6}$$

TYPES OF MATRICES

- *Unsaturated matrix* is a matrix in which the number of tests $N > M$ (number of coefficients in the model). In this case, all coefficients of the regression equation can be determined and at least one order remains free for verification of the model adequacy.
- *Saturated matrix* is the matrix when $N = M$. All coefficients of the regression equation can be determined but the adequacy of the model cannot be verified.
- *Supersaturated matrix* is a matrix in which $N < M$. Only some of the coefficients of the regression equation can be determined and the adequacy of the model cannot be verified.

STAGES OF DERIVATION OF THE MODEL

1. Calculation of coefficients
   Calculation of coefficients is realized by the following formulas:

$$b_0 = \frac{1}{N} \sum_{j=1}^{N} \overline{y}_j \qquad \text{(for zero coefficient)} \tag{II.6.7}$$

$$b_i = \frac{1}{N} \sum_{j=1}^{N} x_{ij} \overline{y}_j \qquad \text{(for linear coefficients)} \tag{II.6.8}$$

$$b_{il} = \frac{1}{N} \sum_{j=1}^{N} x_{ij} x_{lj} \overline{y}_j \qquad \text{(for double interactions)} \tag{II.6.9}$$

$$b_{ilk} = \frac{1}{N} \sum_{j=1}^{N} x_{ij} x_{lj} x_{kj} \overline{y}_j \qquad \text{(for triple interactions).} \tag{II.6.10}$$

2. Verification for reproducibility of the process
As the verification for reproducibility of the process is a verification of equality of more than two dispersions, it is conducted according to the Cochran criterion:

$$G_R = \frac{S_{\max}^2}{\sum_{j=1}^{N} S_j^2} \tag{II.6.11}$$

The table value is accounted from Appendix 5 under the following conditions $G_T(a, f_1 = n - 1, f_2 = N)$ where n is the number of tests for determination of y_j, respectively S_j^2 $(j = 1, 2, \ldots, N)$. If $G_R \leq G_T$, the dispersions are uniform and the process is reproducible and vice versa, if $G_R > G_T$ the process is irreproducible.
3. Calculation of dispersion of the test
If the process is reproducible, the dispersion of the test is determined as

$$S_y^2 = \frac{1}{N}\sum_{1}^{N} S_j^2. \tag{II.6.12}$$

If the process is irreproducible, the dispersion of the test is determined in the classical manner (the deviation of separate values from the mean value):

$$S_y^2 = \frac{1}{N(n-1)}\sum_{1}^{N}\sum_{1}^{n}\left(y_{uj} - \bar{y}_j\right)^2. \tag{II.6.13}$$

4. Determination of dispersion of regression coefficients
It is determined from the dispersion of the test as the product is divided by the number of tests (the combination of input factors) N and the number of tests for determination of the output variable n:

$$S^2(b_i) = \frac{1}{N \cdot n} S_y^2. \tag{II.6.14}$$

5. Verification of significance of calculated regression coefficients
The Student criterion is used. Only those coefficients are significant for which the following is valid:

$$|b_i| \geq t_T \cdot S(b_i), \tag{II.6.15}$$

as $t_T(\alpha = 0.05, f = N(n-1))$.
6. Registration of the derived model
The regression coefficients for which condition (II.6.15) is not fulfilled are ignored and the obtained model is registered.

7. Verification of adequacy of the model
   For evaluation of the deviations of values calculated by the model from the real (measured) values, the variance of adequacy is calculated:

$$S_{ad}^2 = \frac{n}{N-M}\sum_{j=1}^{N}\left(\hat{y}_j - \bar{y}_j\right)^2, \tag{II.6.16}$$

where M is the number of significant coefficient (after the elimination of the insignificant),
$\hat{y}_j$ is the value calculated by the model,
$\bar{y}_j$ is the mean value from the results measured for each combination of input factors.

The calculated value of the Fisher's criterion is determined as the dispersion of adequacy S_{ad}^2 is compared to the dispersion of the test S_y^2:

$$F_R = \frac{S_{ad}^2}{S_y^2}. \tag{II.6.17}$$

The table value of Fisher's distribution is determined under the following conditions: $F_T\,(a = 0.05, f_1 = N - M, f_2 = N - n)$. If $F_R \leq F_T$, the model is adequate. Otherwise it is inadequate and cannot be applied.

Example

A regression equation has to be derived for the influence of three input factors on an output parameter. The natural values of the levels of the factors and the variance interval are given in Table II.6.3.

Table II.6.3. Natural values of the levels of the factors and variance interval

Levels of the factors	x_1	x_2	x_3
$X_{0,i}$—zero level	550	3	10
J_i—variance interval	50	1	5
$X_{u,i}$—upper level	600	4	15
$X_{l,i}$—lower level	500	2	5

A FFE is conducted with $2^3 = 8$ tests as a design of the experiment has been prepared, corresponding to the one in Table II.6.2. Table II.6.4 is supplemented as the double interactions and the triple interaction are added.

(Continued)

(*Continued*)

Table II.6.4. FFE table for the example

Number	x_0	x_1	x_2	x_3	x_1x_2	x_1x_3	x_2x_3	$x_1x_2x_3$	Randomization			y_{uj}			$\bar{y}_j$	S_j^2	$\hat{y}_j$	$(\bar{y}_j - \hat{y}_j)^2$
1	2	3	4	5	6	7	8	9	10	11	12	13	14	15	16	17	18	19
1	+	–	–	–	+	+	+	–	1	17	9	8	11	8	9	3	9	0
2	+	+	–	–	–	–	+	+	16	10	18	13	15	11	13	4	13	0
3	+	–	+	–	–	+	–	+	8	19	2	14	16	12	14	4	14	0
4	+	+	+	–	+	–	–	–	20	11	15	20	18	16	18	4	18	0
5	+	–	–	+	+	–	–	+	3	24	7	16	15	14	15	1	15.5	0.25
6	+	+	–	+	–	+	–	–	14	12	21	22	20	18	20	4	19.5	0.25
7	+	–	+	+	–	–	+	–	6	23	4	17	16	15	16	1	15.5	0.25
8	+	+	+	+	+	+	+	+	22	5	13	21	19	17	19	4	19.5	0.25
														$\Sigma =$	124	25		1

For elimination of random influences, three measurements of the output parameter are made. The order of conducting the tests is determined on a random basis and it is given in the column "Randomization." The results of the conducted tests are presented in columns 13–15. The mean value $\overline{y}_j$ and the variance S_j^2 for each combination of the input factors is given in columns 16 and 17, respectively.

CALCULATION OF COEFFICIENTS

$$b_0 = \frac{1}{N}\sum_{j=1}^{N} \overline{y}_j = \frac{124}{8} = 15.5;$$

$$b_1 = \frac{1}{N}\sum_{j=1}^{N} x_{1j}\overline{y}_j = \frac{-9+13-14+18-15+20-16+19}{8} = 2,$$

$$b_2 = \frac{1}{N}\sum_{j=1}^{N} x_{2j}\overline{y}_j = \frac{-9-13+14+18-15-20+16+19}{8} = 1.25,$$

$$b_3 = \frac{1}{N}\sum_{j=1}^{N} x_{3j}\overline{y}_j = \frac{-9-13-14-18+15+20+16+19}{8} = 2,$$

$$b_{12} = \frac{1}{N}\sum_{j=1}^{N} x_{1j}x_{2j}\overline{y}_j = \frac{+9-13-14+18+15-20-16+19}{8} = -0.25,$$

$$b_{13} = \frac{1}{N}\sum_{j=1}^{N} x_{1j}x_{3j}\overline{y}_j = \frac{+9-13+14-18-15+20-16+19}{8} = 0,$$

$$b_{23} = \frac{1}{N}\sum_{j=1}^{N} x_{2j}x_{3j}\overline{y}_j = \frac{+9+13-14-18-15-20+16+19}{8} = -1.25,$$

$$b_{123} = \frac{1}{N}\sum_{j=1}^{N} x_{1j}x_{2j}x_{3j}\overline{y}_j = \frac{-9+13+14-18+15-20-16+19}{8} = -0.25.$$

VERIFICATION OF REPRODUCIBILITY OF THE PROCESS

It is done with the Cochran criterion:

$$G_R = \frac{S_{\max}^2}{\sum_{j=1}^{N} S_j^2} = \frac{4}{25} = 0.16.$$

The table value is accounted from Appendix 5 under the following conditions G_T ($a = 0.05$, $f_1 = 3-1 = 2, f_2 = 8) = 0.506$. As $G_R \le G_T$, variances are uniform and the process is reproducible.

(*Continued*)

(Continued)

CALCULATION OF THE TEST VARIANCE

The process is reproducible, it is determined with:

$$S_y^2 = \frac{1}{N}\sum_{1}^{N} S_j^2 = \frac{25}{8} = 3.125.$$

DETERMINATION OF THE VARIANCE OF REGRESSION COEFFICIENTS

$$S^2(b_i) = \frac{1}{N \cdot n} S_y^2 = \frac{3.125}{24} = 0.13.$$

VERIFICATION OF THE SIGNIFICANCE OF REGRESSION COEFFICIENTS

From Appendix 1, the value of Student's distribution is accounted under the following conditions $t_T(\alpha = 0.05, f = 8 \times (3-1) = 16) = 2.12$. Significant are those coefficients that are higher in an absolute value than:

$$t_T \cdot S(b_i) = 2.12 \times \sqrt{0.13} = 0.765,$$

that is, the coefficients b_0, b_1, b_2, b_3, and b_{23}. The rest of them should be ignored.

REGISTRATION OF THE DERIVED MODEL

After the elimination of the insignificant coefficients, the model takes the following form:

$$\hat{y} = 15.5 + 2x_1 + 1.25x_2 + 2x_3 - 1.25x_2x_3.$$

CALCULATION OF THE VALUES OF THE OUTPUT VARIABLE ON THE MODEL

The values of $\hat{y}_j$ are calculated, x_1, x_2, x_3 and x_2x_3 are replaced with values (+1) or (−1) in the model depending on their values for the corresponding test. The obtained calculated values are shown in column 18 of Table II.6.4.

VERIFICATION OF THE MODEL ADEQUACY

The adequacy dispersion is calculated from the measured and calculated values using formula (II.6.16) as it is taken into account that the significant coefficients in the model are $M = 5$:

$$S_{ad}^2 = \frac{n}{N-M}\sum_{j=1}^{N}\left(\hat{y}_j - \overline{y}_j\right)^2 = \frac{3}{8-5} \times 1 = 1.$$

As the adequacy dispersion is lower than the test variance, the calculated criterion of Fisher is calculated as

$$F_R = \frac{S_y^2}{S_{ad}^2} = \frac{3.125}{1} = 3.125.$$

The tabular value of Fisher distribution is ($a = 0.05$, $f_1 = 8 - 5 = 3$, $f_2 = 8 \times (3 - 1) = 16$) = 3.24. As $F_R < F_T$ the model is adequate.

II.7. Fractional Factorial Experiment

In case of a large number of input variables, the use of a full factorial experiment is practically impossible. For example, with a number of factors $m = 5$, the number of tests that have to be conducted in order for all coefficients to be determined is $N = 32$ and with $m = 10$ the number of tests is $N = 1024$. However, practice shows that in most cases the coefficients before the interactions, especially the ones with a big number of factors, are insignificant.

Fractional factorial experiment (FrFE) is used when the aim is only for the coefficients before the separate factors to be found without determining the coefficients before the different interactions. It consists of a replacement of some interactions (especially the ones with many factors) with new factors. In this case, all characteristics of a full factorial experiment remain the same. Usually a fractional factorial experiment is applied for a number of input factors $m > 4$.

Fractional replica is a design containing a part of the tests of a full factorial experiment. *The degree of fractionality* is determined by the number of interactions replaced with a new factor. The parities obtained are called *generating correlations*. For example, if the evaluations of the coefficients in the following model have to be found:

$$\hat{y} = b_0 + b_1x_1 + b_2x_2 + b_3x_3, \tag{II.7.1}$$

four tests are necessary for their determination because the coefficients to be found are 4. The factor x_3 can be obtained by the generating correlation (GC):

$$x_3 = x_1x_2, \tag{II.7.2}$$

and its coefficient b_3 is a mixed evaluation for β_3 and β_{12}:

$$b_3 \rightarrow \beta_3 + \beta_{12}. \tag{II.7.3}$$

When the area of variance of the factors is small the model can be considered linear and the interactions practically do not exist ($\beta_{12} = 0$), that is, the coefficient b_3 will evaluate only β_3.

The *number of tests* in fractional replicas is determined when the number of tests of the full factorial experiment is divided by one of the orders of the number 2:

$$N = 2^{m-p}, \tag{II.7.4}$$

where p is the number of generating correlations or the degree of fractionality. The number of tests for the most commonly used replicas is

- 1/2 replica (semireplica) $N_{FrFE} = \frac{2^m}{2^1} = 2^{m-1}$
- 1/4 replica (quarter-replica) $N_{FrFE} = \frac{2^m}{2^2} = 2^{m-2}$
- 1/8 replica (one-eighth–replica) $N_{FrFE} = \frac{2^m}{2^3} = 2^{m-3}$

STAGES OF CONSTRUCTION OF A FRACTIONAL FACTORIAL EXPERIMENT

1. The minimum number of tests for determination of the coefficients is set:

$$N_{\min} \geq m + 2, \qquad \text{(II.7.5)}$$

   where m is the number of factors. The additional tests are two (one for the zero coefficient and one for verification of the model adequacy).
2. The level of fractionality p is set so that:

$$N_{FrFE} > 2^{m-p}. \qquad \text{(II.7.6)}$$

3. $(m-p)$ main factors are chosen, the other p are additional and will be introduced with the help of generating correlations.
4. Full factorial experiment is constructed for the main factors.
5. The generating correlations are given as the additional factors are equalized with the products of highest order of the main factors.
6. The table of the design is complemented with the generating correlations.

The so-called *determining contrasts* (DC) are used for clarifying the method of mixing the evaluations. Determining contrast is obtained when the two sides of the generating correlation are multiplied by the additional factor. For the example above:

$$DC = x_3^2 = x_1 x_2 x_3. \qquad \text{(II.7.7)}$$

When the generating correlations, respectively the determining contrasts are more than one, *generalized determining contrast* (GDC) is set. It is all the possible combinations of the determining contrasts. In order to determine which coefficients are mixed, the determining contrast, respectively the generalized determining contrast, is multiplied by each member of the model. The calculation of coefficients and the verification of their significance and adequacy of the model are the same as in the full factorial model.

Example

A design of a fractional factorial experiment should be composed with the help of which the strength of the sized warp yarn should be examined depending on seven input factors:

x_1—percentage of size (%)
x_2—temperature of size (°C)
x_3—concentration of size (%)
x_4—pressure of squeezing rollers (Pa)
x_5—sizing speed (m/min)
x_6—drying time (min)
x_7—temperature of drying of the warp (°C).

The method of mixing the coefficients should be determined.

DETERMINATION OF THE MINIMUM NUMBER OF TESTS FOR DERIVING A LINEAR MODEL

$N > 7 + 2 = 9$
$N \Re{}_{|}^{|}\,^{2}$ —insufficient number of tests
$N = 2^3 = 8 < 9$—insufficient number of tests
$N = 2^4 = 2^{7-3} = 16 > 9$—the number of tests is sufficient as the main factors are 4 and the additional $(m - p) = 3$

CHOICE OF MAIN AND ADDITIONAL FACTORS

We choose x_1, x_2, x_3, and x_4 to be the main factors.
Possible combinations of the main factors are determined:

$$x_1x_2; \quad x_1x_3; \quad x_1x_4; \quad x_2x_3; \quad x_2x_4; \quad x_3x_4;$$
$$x_1x_2x_3; \quad x_1x_2x_4; \quad x_1x_3x_4; \quad x_2x_3x_4; \quad x_1x_2x_3x_4.$$

The last three combinations are chosen to be the generating correlations:

$$x_5 = x_1x_3x_4; \qquad x_6 = x_2x_3x_4; \qquad x_7 = x_1x_2x_3x_4.$$

COMPOSING THE DESIGN OF THE EXPERIMENT

The plan of the experiment is presented in Table II.7.1.

(*Continued*)

(*Continued*)

Table II.7.1. Plan of the experiment for the example

Number	x_1	x_2	x_3	x_4	$x_5 = x_1x_3x_4$	$x_6 = x_2x_3x_4$	$x_7 = x_1x_2x_3x_4$	y
1	+	+	+	+	+	+	+	y_1
2	−	+	+	+	−	+	−	y_2
3	+	−	+	+	+	−	−	y_3
4	−	−	+	+	−	−	+	y_4
5	+	+	−	+	−	−	−	y_5
6	−	+	−	+	+	−	+	y_6
7	+	−	−	+	−	+	+	y_7
8	−	−	−	+	+	+	−	y_8
9	+	+	+	−	−	−	−	y_9
10	−	+	+	−	+	−	+	y_{10}
11	+	−	+	−	−	+	+	y_{11}
12	−	−	+	−	+	+	−	y_{12}
13	+	+	−	−	+	+	+	y_{13}
14	−	+	−	−	−	+	−	y_{14}
15	+	−	−	−	+	−	−	y_{15}
16	−	−	−	−	−	−	+	y_{16}

SETTING THE DETERMINING CONTRASTS

$$DC_1 = x_5^2 = x_1x_3x_4x_5 = 1$$
$$DC_2 = x_6^2 = x_2x_3x_4x_6 = 1$$
$$DC_3 = x_7^2 = x_1x_2x_3x_4x_7 = 1.$$

SETTING THE GENERALIZED DETERMINING CONTRAST

$$GDC = DC_1 = DC_2 = DC_3 = DC_1 \cdot DC_2 = DC_1 \cdot DC_3 = DC_2 \cdot DC_3$$
$$= DC_1 \cdot DC_2 \cdot DC_3 = 1.$$

When we take into account that in the presence of a factor in the second order the GDC always is one, we get the following:

$$GDC = x_1x_3x_4x_5 = x_2x_3x_4x_6 = x_1x_2x_3x_4x_7 = x_1x_2x_5x_6 = x_2x_5x_7 = x_1x_6x_7$$
$$= x_3x_4x_5x_6x_7$$

MIXING THE COEFFICIENTS

In order to determine for which coefficients the coefficient b_1 is a mixed evaluation, the generalized determining contrast is multiplied by x_1.

$$x_1 \cdot DC_1 = x_1x_1x_3x_4x_5 = x_3x_4x_5,$$
$$x_1 \cdot DC_2 = x_1x_2x_3x_4x_6,$$
$$x_1 \cdot DC_3 = x_1x_1x_2x_3x_4x_7 = x_2x_3x_4x_7,$$
$$x_1 \cdot DC_1 \cdot DC_2 = x_1x_1x_2x_5x_6 = x_2x_5x_6,$$
$$x_1 \cdot DC_1 \cdot DC_3 = x_1x_2x_5x_7,$$
$$x_1 \cdot DC_2 \cdot DC_3 = x_1x_1x_6x_7 = x_6x_7,$$
$$x_1 \cdot DC_1 \cdot DC_2 \cdot DC_3 = x_1x_3x_4x_5x_6x_7.$$

The evaluation b_1 evaluates the coefficient β_1 and the coefficients before the products obtained above:

$$b_1 \to \beta_1 + \beta_{345} + \beta_{12346} + \beta_{2347} + \beta_{256} + \beta_{1257} + \beta_{67} + \beta_{134567}.$$

Mixing shows that alternation of (+) and (–) in the columns of the design corresponding to these products is the same, that is, only one coefficient can be determined from the specified ones and for our task this will be the coefficient before the factor x_1.

The evaluation b_2 evaluates the coefficient β_2 and the coefficients:

$$b_2 \to \beta_2 + \beta_{12345} + \beta_{346} + \beta_{1347} + \beta_{156} + \beta_{57} + \beta_{1267} + \beta_{234567},$$

which are determined when the generalized determining contrast is multiplied by x_1.

Mixings for the other linear coefficients are determined in an analogical way:

$$b_3 \to \beta_3 + \beta_{145} + \beta_{246} + \beta_{1247} + \beta_{12356} + \beta_{2357} + \beta_{1367} + \beta_{4567},$$
$$b_4 \to \beta_4 + \beta_{135} + \beta_{236} + \beta_{1237} + \beta_{12456} + \beta_{2457} + \beta_{1467} + \beta_{3567},$$
$$b_5 \to \beta_5 + \beta_{134} + \beta_{23456} + \beta_{123457} + \beta_{126} + \beta_{27} + \beta_{1567} + \beta_{3467},$$
$$b_6 \to \beta_6 + \beta_{13456} + \beta_{234} + \beta_{123467} + \beta_{125} + \beta_{2567} + \beta_{17} + \beta_{3457},$$
$$b_7 \to \beta_7 + \beta_{13457} + \beta_{23467} + \beta_{1234} + \beta_{12567} + \beta_{25} + \beta_{16} + \beta_{3456}.$$

The mixings of the coefficients for the interactions that are not replaced with generating correlations can also be determined. The number of coefficients before the interactions that can be determined is: $N - k - 2 = 16 - 7 - 2 = 7$. Let's assume that these are the first seven of the combinations of interactions listed above. The last three are used for generating correlations and only the coefficient before the eighth interaction $x_1x_2x_4$ cannot be determined. Theoretically, it can be calculated but in this case no degree of freedom will remain for verification of the model.

If the coefficient b_{12} is determined it will evaluate:

$$b_{12} \to \beta_{12} + \beta_{2345} + \beta_{1346} + \beta_{347} + \beta_{56} + \beta_{157} + \beta_{267} + \beta_{1234567}.$$

The indices are obtained when the generalized determining contrast is multiplied by x_1x_2. The mixings of the coefficients x_1x_3, x_1x_4, x_2x_3, x_2x_4, x_3x_4, $x_1x_2x_3$ in an analogical way.

These mixings are determined rarely as the practice is to determine only the linear coefficients.

II.8. Statistical Methods for Movement to an Optimal Area

The aim of the production processes and objects investigation, most often, is their optimization, that is, finding values of the factors which give the best result. The desired result, depending on the indicator, can be both the minimum and the maximum value of the target function. This is why the task of optimization is usually regarded as a task for finding an extremum. When the extremum of the target function is sought by means of an experiment, the following characteristics should be considered:

1. Because of the presence of disturbing influences, the measured value of the optimization parameter is a random variable.
2. Often, the experimenter does not know even the approximate location of the extremum.

The first characteristic determines the statistical nature of the methods for finding an extremum. The second one shows that if the experimenter decides to perform a mathematical description of the object in the whole area of factor change and then looks for the extremum, it will be very difficult for him. The bigger the factor change area is, the more complex the center of the experiment is. Therefore, the equation that describes it is more complex. This is why with experimental optimization, the movement to the extremum is done by successively studying the target surface with experiments and organizing steps towards improvement of the target function.

There are a number of methods for movement to the extremum which mainly differ in the direction of the motion towards it which is hard to determine. Some of the most common methods are the Box–Wilson and the simplex methods.

BOX–WILSON METHOD

The method is also known as the steep ascend/descend method. Box and Wilson suggested finding the optimal conditions to be carried out in two stages. In the first stage, there are first-order local mathematical descriptions and movements in the direction of the gradient that are carried out to reach the extremum area. In the second stage, experimental plans are constructed which allow a more detailed study of the optimal area and they will be discussed in Section II.9.

CONDITION FOR APPLICATION OF THE METHOD

The method is applicable if the target function is a linear regression equation, with fixed control parameters worked out through a full factorial experiment or a fractional factorial experiment, with the general form:

$$\hat{y} = b_o + b_1 x_1 + b_2 x_2 + \ldots + b_n x_n \qquad \text{(II.8.1)}$$

and at least one significant linear regression coefficient b_i.

METHOD PRINCIPLE

If a series of experiments is performed in which for each subsequent version the factors x_i vary proportionally to the product of the regression coefficient of the given factor and the range of its variation ($b_i \cdot J_i$), then such motion is the shortest way to the optimal area. The motion in the chosen direction continues until the optimal value is reached. After that, a new planned experiment is carried out with its center in the extremum. If at least one of the coefficients b_i is different from zero, a new motion direction is sought. The method is represented graphically in Figure II.8.1. The main phases are:

- I is an experiment (FFT or FrFT) in the starting area;
- movement to the optimum area following the first gradient direction;
- II is an experiment (FFT or FrFT) in the area, where the linear model deviates from the real experiment;
- movement to the optimum area following the second gradient direction;
- III is an experiment for obtaining of a more precise second-order model.

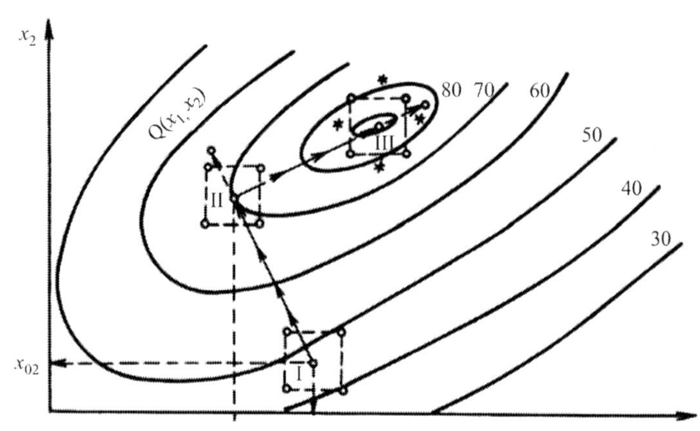

Figure II.8.1. Graphical presentation of the method of Box and Wilson.

METHOD APPLICATION CONDITIONS

1. Process manageability—ability to change the factors at different levels.
2. Reproducibility of the process—homogeneity of dispersions at the various points of the experiment.
3. Stationarity of the process—at the same point, regardless of the time, the dispersion is constant.

DISADVANTAGES OF THE METHOD

1. Only one output parameter can be examined.
2. The information from the previous experiment cannot be used for the next one.

APPLICATION

This method is suitable for quick identification of an optimal regime under laboratory conditions. It will be clarified with an example.

Example

The optimal temperature t (°C) and concentration C (%) must be determined when dyeing of fabric to obtain a maximum degree of fixation of the dye Q (%). The dyeing temperature is marked with x_1 and the concentration of the dye solution with x_2.

The main factor levels, the interval of variation, as well as their upper and lower levels are given in Table II.8.1.

Table II.8.1. Main factor levels, interval of variation, upper and lower levels

Factors	x_1	x_2
$x_{o,i}$	80	32.6
J_i	5	1
$x_{u,i}$	85	33.6
$x_{l,i}$	75	31.6

To work out a model, a full factorial experiment has been conducted with a number of tests $N = 2^2 = 4$. For each combination of factors, two tests are conducted at random. The values of the output parameter y_1 and y_2 are determined. The following statistical assessments are defined as well—average values $y_{j,e}$ and variances S_j^2. The results are presented in Table II.8.2.

(Continued)

(*Continued*)

Table II.8.2. Experimental data and calculations for obtaining of the first model

Number	x_1	x_2	Randomization		$y_{j,1}$	$y_{j,2}$	y	S_j^2	$y_{j,R}$	$(y_{j,e}-y_{j,R})^2$
1	−1	−1	5	3	83.5	84.5	84	0.5	84.05	0.0025
2	1	−1	1	7	88	87	87.5	0.5	87.45	0.0025
3	−1	1	8	6	81	82.4	81.7	0.98	81.65	0.0025
4	1	1	4	2	85.5	84.5	85	0.5	85.05	0.0025
							Σ =	2.48		0.01

Based on the above data, the coefficients of the regression equation are defined:

$$b_0 = 84 + 87.5 + 81.7 + 85 = 84.55,$$

$$b_1 = -84 + 87.5 - 81.7 + 85 = 1.7,$$

$$b_2 = 84 - 87.5 + 81.7 - 85 = -1.2.$$

A check for homogeneity of the variances is performed. Since the variances are more than two, the Cochran criterion is applied. Its expected value is determined according to formula (II.6.11):

$$G_R = \frac{S_{\max}^2}{\sum_{j=1}^{k} S_j^2} = \frac{0.98}{2.48} = 0.395.$$

The table value G_T ($a, f_1 = n - 1, f_2 = k$) is accounted for in Appendix 5 as n is the number of experiments in the groups and k is the number of groups. The reported value is G_T (0.05, $f_1 = 2 - 1 = 1, f_2 = 4$) = 0.907. Since $G_R \le G_T$ it is accepted that the variances are homogenous and the process is reproducible.

The reproducibility of the process is the reason the variance of the test is defined as an average value of variances of the different tests [from formula (II.6.12)]:

$$S_y^2 = \frac{1}{N}\sum_{1}^{N} S_j^2 = \frac{1}{4}(0.5 + 0.5 + 0.98 + 0.5) = 0.62.$$

Dispersion of regression coefficients is determined according to formula (II.6.14):

$$S^2(b_i) = \frac{1}{N.n} S_y^2 = \frac{1}{4 \times 2} 0.62 = 0.078.$$

The significance check of the calculated regression coefficients is performed by the Student's t-criterion, as the variable $t_T \cdot \frac{S(b_i)}{\sqrt{n}}$ is determined. When t_T(a = 0.05, $f = N(n-1)$ $= 4 \times (2-1) = 4) = 0.276$, those coefficients are significant, for which:

$$|b_i| \geq t_T \cdot \frac{S(b_i)}{\sqrt{n}} = 2.776 \times \sqrt{\frac{0.078}{2}} = 0.548.$$

that is, both coefficients b_1 and b_2 are significant.

The adequacy of the model, $y = 84.55 + 1.7x_1 - 1.2x_2$, is checked by first calculating the expected values according to the model $y_{j,R}$ (penultimate column of Table II.8.2) and then calculating the squares of the deviations of the expected values from the experimental data (the last column of Table II.8.2). By the sum of the deviations of formula (II.6.16), the dispersion of adequacy is defined:

$$S_{ad}^2 = \frac{n}{N-M} \sum_{j=1}^{N} \left(y_{j,e} - y_{j,R}\right)^2 = \frac{2}{4-3} 0.01 = 0.02,$$

where $M = 3$ is the number of significant coefficients. The dispersion of adequacy value is lower than the one of the dispersion of the test, therefore the expected value of the Fisher criterion is defined as

$$F_R = \frac{S_y^2}{S_{ad}^2} = \frac{0.62}{0.02} = 31.$$

The table value of the Fisher's distribution is determined under the following conditions: F_T $(a = 0.05, f_1 = N - n = 4 - 2 = 2, f_2 = N - M = 4 - 3 = 1) = 199.5$. $F_R \leq F_T$, that is, the model is adequate.

As at least one regression coefficient is significant and all the other requirements for the application of the method are met, we proceed to the procedure for movement to the extremal area. The work algorithm is the following:

1. A base factor is chosen.
   The basic factor is chosen to be the one for which the product $|b_i \cdot J_i|$ is greater.

   $$|b_1 \cdot J_1| = |1.7 \times 5| = 8.5 \qquad |b_2 \cdot J_2| = |-1.2 \times 1| = 1.2$$

   As $|b_1 \cdot J_1| > |b_2 \cdot J_2|$, then for base factor we choose the factor x_1.
2. A base step is chosen and the variation step of the other factor.
   Most often, the variation interval of the base factor is chosen, that is: $S_B = S_1 = 5$. Based on this, the variation step of the factor x_2 is calculated:

$$S_2 = \frac{b_2 \cdot J_2}{b_B \cdot J_B} S_B = \frac{-1.2 \times 1}{1.7 \times 5} 5 = -0.7. \qquad \text{(II.8.2)}$$

3. The expected values are calculated and compared to the experimental ones. The expected values are calculated by the formula:

$$y_R = b_0 + b_1 \frac{S_1}{J_1} + b_2 \frac{S_2}{J_2}. \tag{II.8.3}$$

When the tendency of variation of the expected values matches with the ones of the experimental values, one experiment may be skipped and only an imaginary test may be performed. In Table II.8.3 the levels of the factors x_1 and x_2 are given, as well as the calculated (expected) y_R, and the experimental results y_E.

Table II.8.3. Computational table for moving to the optimal area

Factors	x_1	x_2	y_R	y_E
$b_i J_i$	8.5	–1.2		
S_i	5	–0.7		
Test 0 (main level)	80	32.6	84.55	86.6
Test 1	85	31.9	87.09	87.5
Test 2 (imaginary)	90	31.2	89.63	—
Test 3	95	30.5	92.17	91.7
Test 4 (imaginary)	100	29.8	94.71	– (92.2) ←
Test 5	105	29.1	97.25	92
Test 6	110	28.4	99.79	83.1

The zero test concurs with the center of the experiment. The expected value for it is $y_R = 84.55$, and the value determined through an experiment $y_E = 86.6$. Test 1 corresponds to levels of the factors $x_{1,1} = x_{0,1} + S_1 = 80 + 5 = 85$ and $x_{1,2} = x_{0,2} + S_2 = 32.6 - 0.7 = 31.9$. The expected value is

$$y_R = 84.55 + 1.7\frac{5}{5} - 1.2\frac{-0.7}{1} = 87.09.$$

Experimentally, the determined value $y_E = 87.5$ also increases. This fact allows the next test to be only "imaginary."

The expected value for Test 2 is determined:

$$y_R = 84.55 + 1.7\frac{2 \times 5}{5} - 1.2\frac{2 \times (-0.7)}{1} = 89.63.$$

Test 3 concurs at levels of the factors $x_{3,1} = x_{2,1} + S_1 = 90 + 5 = 95$ and $x_{3,2} = x_{2,2} + S_2 = 31.2 - 0.7 = 30.5$.

The expected value is determined:

$$y_R = 84.55 + 1.7\frac{3\times 5}{5} - 1.2\frac{3\times(-0.7)}{1} = 92.17,$$

as well as the experimental value $y_E = 91.7$. The tendency for increase in the output parameter is valid, both, for the expected and for the experimental values. This is why Test 3 is "imaginary." The expected value is

$$y_R = 84.55 + 1.7\frac{4\times 5}{5} - 1.2\frac{4\times(-0.7)}{1} = 94.71.$$

Test 5 is carried out at factor levels $x_{5,1} = x_{4,1} + S_1 = 100 + 5 = 105$ and $x_{5,2} = x_{4,2} + S_2 = 29.8 - 0.7 = 29.1$.

The expected value is

$$y_R = 84.55 + 1.7\frac{5\times 5}{5} - 1.2\frac{5\times(-0.7)}{1} = 97.25,$$

and the experimentally established one is $y_E = 92$.

The output parameter calculated by the model and the experimentally determined one increase but the difference between them is significant. Therefore, for Test 6 it is calculated and determined experimentally. The expected value for Test 6 is

$$y_R = 84.55 + 1.7\frac{6\times 5}{5} - 1.2\frac{6\times(-0.7)}{1} = 99.79,$$

and the experimental one is $y_E = 83.1$.

There is a discrepancy between the real and the experimental values which shows that the extremum has been reached and bypassed (see Test III in Figure II.8.1). In this case, we should go back and actually perform the imaginary Test 4 to determine the extremal point.

The experimental value in Test 4 is 92.2. Since this is the highest value, a new full factorial experiment is given in Table II.8.4 with its center in it.

Table II.8.4. Main factor levels, interval of variation, upper and lower levels for the new FFE

Factors	x_1	x_2
$x_{o,i}$	100	29.8
J_i	5	1
$x_{u,i}$	105	30.8
$x_{l,i}$	95	28.8

The second series of tests may be carried out without retests because the variance S_y^2 is known. Results from the experiment are shown in Table II.8.5.

Table II.8.5. Experimental date for the FFE

Tests	x_1	x_2	y
1	–	–	88
2	+	–	90
3	–	+	92.5
4	+	+	94

The new coefficients of the regression model are determined:

$$b_0 = 91.1; \qquad b_1 = 0.88; \qquad b_2 = 2.12.$$

$|b_i| \geq 0.548$, that is, the condition for application of the Box–Wilson method is present. The new model is:

$$y = 91.1 + 0.88x_1 + 2.12x_2.$$

The same algorithm repeats itself.

1. $|b_1 \cdot J_1| = |0.88 \times 5| = 4.4 \qquad |b_2 \cdot J_2| = |2.2 \times 1| = 2.2$

   Since $|b_1 \cdot J_1| > |b_2 \cdot J_2|$, we choose the factor x_1 for basic factor.

2. $S_B = S_1 = 5 \qquad S_2 = \dfrac{b_2 \cdot J_2}{b_B \cdot J_B} S_B = 2.4$

3. The expected values of the output parameter are calculated and its experimental values are determined (Table II.8.6).

Table II.8.6. Computational table for moving to the optimal area according to the new direction

Factors	x_1	x_2	y_R	y_E
$b_i J_i$	4.4	2.12		
S_i	5	2.4		
Test 0 (main level)	100	29.8	91.1	92.2
Test 1	105	32.2	97.1	94
Test 2	110	34.6	103.0	97.1
Test 3	115	37	109.0	94.2
Test 4	120	39.4	115.0	78

The levels $x_1 = 110$ and $x_2 = 34.6$ at which the optimal value of the output parameter is obtained are chosen for main levels for planning of an experiment to receive a model of second order.

SIMPLEX METHOD

A characteristic of the simplex method for optimization is the simultaneous study of the surface of response and the movement towards the optimum. The experiments are carried out at the top of the simplex.

A *simplex* is a convex polyhedron in the n-dimensional space with $(n+1)$ number of apices. Examples of simplices in the zero-, one-, two-, and three-dimensional space are presented in Figure II.8.2.

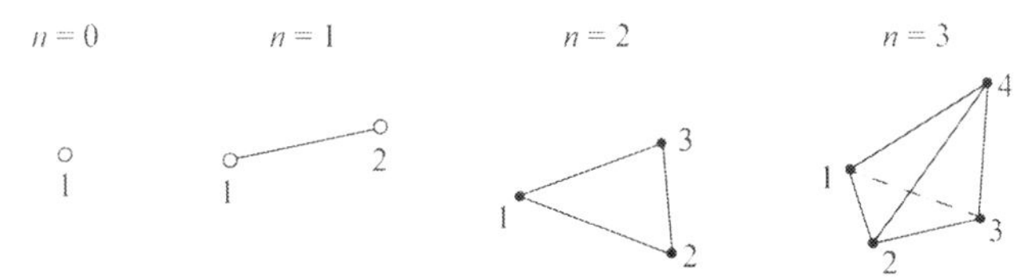

Figure II.8.2. Examples of simplices.

A *regular simplex* is a simplex with regular distances between its apices.

SIMPLEX PROPERTY

From each simplex a new one can be obtained by eliminating one of its points and adding a new one, which is symmetric in relation to the opposite side (Figure II.8.3).

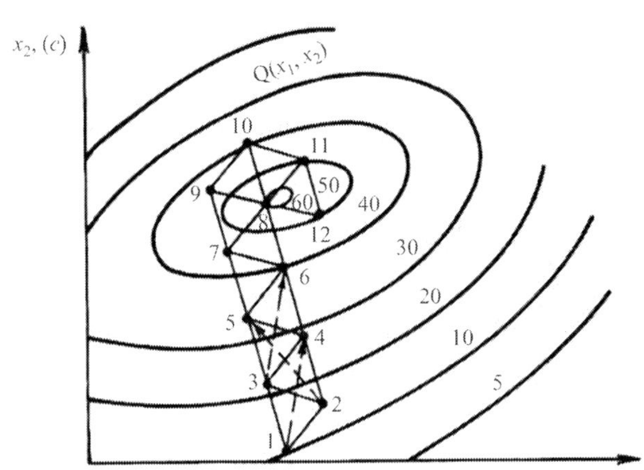

Figure II.8.3. "Rolling" of the simplex.

CRITERION FOR REACHING THE OPTIMAL AREA

A criterion for reaching the optimal area is the wedging or looping of the simplex which finds expression in its rotation around one point. It should be noted here that the wedging may occur also when errors are made in determining the target function. Therefore a retest is made at the rotation point. After the retest the point will either be eliminated (if it was incorrect) or it will be confirmed that it is an optimum (if it has been correctly determined).

SPECIFYING THE OPTIMAL AREA

Specifying the location of the optimal area is made by decreasing the size of the simplex, for example, when we find an opposite point the distance may be halved.

CONSTRUCTION OF THE INITIAL SIMPLEX

An initial simplex may be constructed in any way but the following two methods have gained the widest currency:

1. A regular simplex with the length of its one side being 1 and situated so that one of its apices lies on one of the coordinate axes and the beginning of the coordinate axis lies in the center of gravity of the simplex (Figure II.8.4).

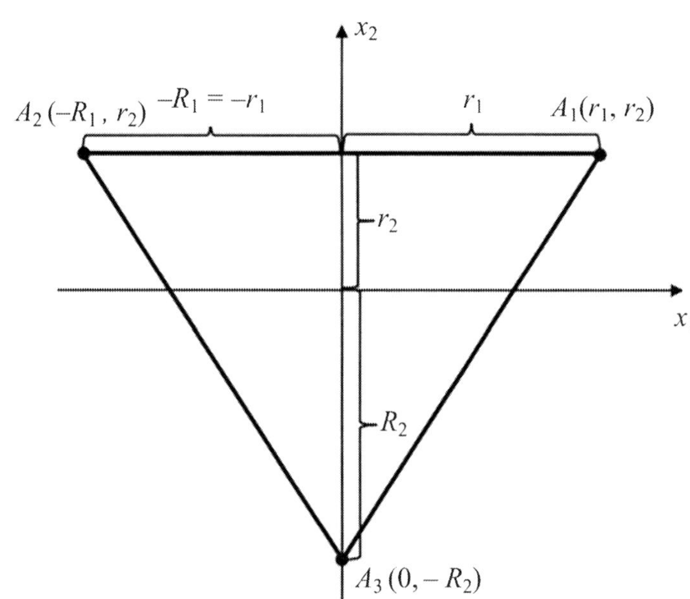

Figure II.8.4. Construction of an initial simplex.

The coordinates of the different points are determined according to Table II.8.7. The values of r_i and R_i are determined as the radii of the circles that inscribe and circumscribe a triangle:

$$r_i = \frac{1}{\sqrt{2 \cdot i(i+1)}} \qquad \text{(II.8.4)}$$

and

$$R_i = \sqrt{\frac{i}{2(i+1)}}. \qquad \text{(II.8.5)}$$

Table II.8.7. Determination of the coordinates of the initial points

	Values of the factor							
Number	x_1	x_2	x_3	...	x_i	...	x_n	y
1	r_1	r_2	r_3	...	r_i	...	r_n	y_1
2	$-R_1 = r_1$	r_2	r_3	...	r_i	...	r_n	y_2
3	0	$-R_2 = 2r_2$	r_3	...	r_i	...	r_n	y_3
...	...	...	...	...	...	...	...	
n	0	0	0	...	$-R_i = i \cdot r_i$	...	r_n	y_n
$n+1$	0	0	0	...	0	...	$-R_n = nr_n$	y_{n+1}

Depending on the number of the input factors, a part of the table is used.

2. A regular simplex with the distance between its apices being 1, situated so that one of its apices lies in the beginning of the coordinate system and the sides that go from it under the same angles with the respective coordinate axes (Figure II.8.5).

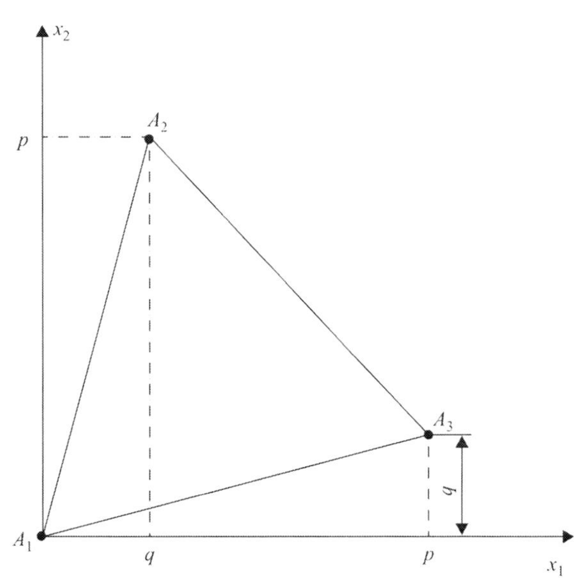

Figure II.8.5. Construction of an initial simplex.

The coordinates of the different points are given in Table II.8.8. The values of the parameters p and q are determined depending on the dimensionality of the space:

$$p = \frac{1}{n\sqrt{2}}\left(n - 1 + \sqrt{n+1}\right) \tag{II.8.6}$$

and

$$q = \frac{1}{n\sqrt{2}}\left(\sqrt{n+1} - 1\right). \tag{II.8.7}$$

Table II.8.8 Determination of the coordinates of the initial points

	Values of the factor							
Number	x_1	x_2	x_3	...	x_i	...	x_n	y
1	0	0	0	...	0	...	0	y_1
2	p	q	q	...	q	...	q	y_2
3	q	p	q	...	q	...	q	y_3
...	...	...	...	...	...	...	...	
n	q	q	q	...	q	...	q	y_n
$n + 1$	q	q	q	...	q	...	p	y_{n+1}

CALCULATING THE VALUES OF THE COORDINATE POINTS

The values of the collimating marks in natural units are calculated as

$$\begin{aligned} x_{11} &= x_{01} + z_{11}J_1 \qquad & x_{21} &= x_{02} + z_{21}J_2 \\ x_{12} &= x_{01} + z_{12}J_1 \qquad & x_{22} &= x_{02} + z_{22}J_2 \\ x_{13} &= x_{01} + z_{13}J_1 \qquad & x_{23} &= x_{02} + z_{23}J_2, \end{aligned} \tag{II.8.8}$$

where x_{01} is the main level of the factor x_1,
x_{02} is main level of the factor x_2,
z_{ii} is value of the matrix in Table II.8.7,
J_1 is interval of variation of the factor x_1,
J_2 is interval of variation of the factor x_2.

CALCULATING THE COORDINATES OF THE MIRRORED POINT APEX

After determining the worst result, the corresponding point is eliminated and the coordinates of the mirrored point are calculated:

$$x_{ig} = \frac{2}{n}\sum_{\substack{j=1 \\ j \neq p}}^{n+1} x_{ij} - x_{ip}, \tag{II.8.9}$$

where p is the index of the eliminated point. The procedure for elimination of tests and determining new experimental points continues until a "rotation" of the simplex is obtained. After specifying the extremal area, a planned experiment of second order is concurred in it, which is described in Section II.9.

Example

Find the area of the extremum of the strength of the yarn (in cN) with change of the twists, factor x_1, and the spindle revolutions, factor x_2.

$$x_{01} = 300 \text{ m}^{-1} \qquad x_{02} = 8000 \text{ min}^{-1}$$

$$J_1 = 20 \text{ m}^{-1} \qquad J_2 = 1000 \text{ m}^{-1}$$

We will determine the coefficients z_{ij} with two factors, according to the first method of locating the starting points, as shown in Table II.8.9.

Table II.8.9. Coordinates of the initial points

z_{ij}	x_1	x_2
1	0.5	0.289
2	–0.5	0.289
3	0	–0.578

DETERMINING THE COORDINATES OF THE STARTING POINTS

$$x_{11} = x_{01} + z_{11}J_1 = 300 + 0.5 \times 20 = 310 \text{ m}^{-1}$$
$$x_{12} = x_{01} + z_{12}J_1 = 300 - 0.5 \times 20 = 290 \text{ m}^{-1}$$
$$x_{13} = x_{01} + z_{13}J_1 = 300 + 0 \times 20 = 300 \text{ m}^{-1}$$
$$x_{21} = x_{02} + z_{21}J_2 = 8000 + 0.289 \times 1000 \approx 8300 \text{ min}^{-1}$$
$$x_{22} = x_{02} + z_{22}J_2 = 8000 + 0.289 \times 1000 \approx 8300 \text{ min}^{-1}$$
$$x_{23} = x_{02} + z_{23}J_2 = 8000 - 0.578 \times 1000 \approx 7400 \text{ min}^{-1}$$

FILLING IN THE SIMPLEX TABLE

To start with, in Table II.8.10, we enter the coordinates of the three points forming the initial simplex S_0. The values of the output parameter y for each of the three points are determined experimentally.

(Continued)

Table II.8.10. Coordinates of the simplices

Number	Factors x_1	x_2	Number of the simplex	y	Number of the eliminated point
1	310	8300	S_0	210	
2	290	8300	S_0	200	
3	300	7400	S_0	150	3
4	300	9200	S_1	230	2
5	320	9200	S_2	250	

From the three results, the most unfavorable is eliminated. Since the target parameter is strength, the lowest value 150 cN (third point) is eliminated.

DETERMINING THE COORDINATES OF AN ADDITIONAL POINT

The following formula is used from formula (II.8.9):

$$x_{1g} = \frac{2}{2}(310 + 290) - 300 = 300\,\text{min}^{-1}.$$

$$x_{2g} = \frac{2}{2}(8300 + 8300) - 7400 = 9200\,\text{min}^{-1}.$$

FORMATION OF A NEW SIMPLEX

An additional fourth point forms simplex S_1. The strength at levels $x_1 = 300\ \text{m}^{-1}$ and $x_2 = 9200\ \text{min}^{-1}$ is determined. The most unfavorable result from simplex S_1 (points 1, 2, and 4) is chosen. Point 2 is eliminated and the coordinates of the new additional point are determined:

$$x_{1g} = \frac{2}{2}(300 + 310) - 290 = 320\ \text{m}^{-1},$$

$$x_{2g} = \frac{2}{2}(8300 + 9200) - 8300 = 9200\,\text{min}^{-1}.$$

A test at levels $x_1 = 320\ \text{m}^{-1}$ and $x_2 = 9200\ \text{min}^{-1}$ is carried out. From simplex S_2 (points 1, 4, and 5), the most undesirable result is determined. Point 1 is eliminated and its opposite point is made. This procedure continues until the optimal area is reached.

II.9. Investigation of the Optimum Area: Composite Designs of Second Order

After determination of the approximate location of the optimum area by some of the methods presented in Section II.8, the next step is its detailed study. The extremum area can be described mathematically with a regression equation. As nonlinearity is strongly expressed, the use of a full factorial experiment or its fractional replicas is unsuitable.

In practice, polynomial models of second order are one of the most often used. They allow for the studied phenomenon to be described in a relatively wide area of variation of the input variables. Their general form is

$$\hat{y} = b_0 + \sum_{i=1}^{m} b_i x_i + \sum_{i<j} b_{ij} x_i x_j + \sum_{i=1}^{m} b_{ii} x_i^2 . \tag{II.9.1}$$

The composite designs of second order consist of the following types of points:

1. Points of full or fractional factorial experiment, the number of which is N_{FE}.
2. Star points which are located on the coordinate axes at equal intervals α from the center of the design, called "star arm." Their number is N_a.
3. Observations in the center of the design which is indicated with c N_0, as for some designs $N_0 = 0$.

The total number of tests is

$$N = N_{FE} + N_a + N_0 = 2^{m-p} + 2m + N_0, \tag{II.9.2}$$

where p is the degree of fractionality (in the full factorial experiment $p = 0$). In Figure II.9.1 the position of the points from the composite designs for (a) 2 and (b) 3 factors is shown.

The first designs of this type are the *orthogonal central composite designs* advanced by Box–Wilson. In these designs the star arm is chosen so that after the transformation of the variables a diagonal information matrix is obtained which simplifies the calculation of regression coefficients. *Rotatable central composite designs* are also applied. They are designed so that they ensure equal dispersion of the forecasted value of the output variable at equal intervals from the center. Later, the so-called *optimal composite designs* were advanced, which ensured

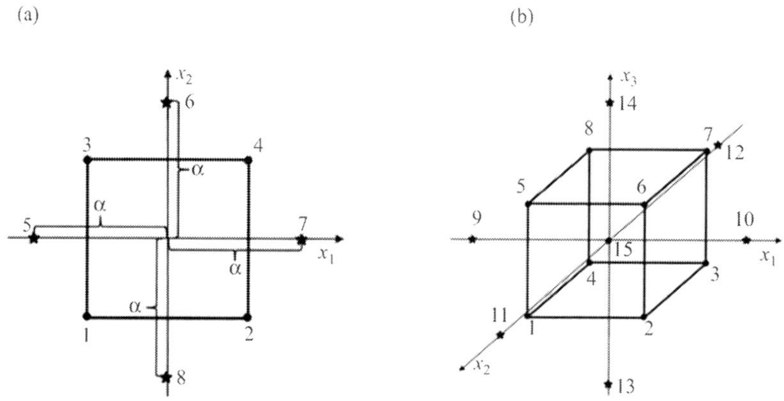

Figure II.9.1. Experimental point for composite designs.

obtaining effective evaluations and minimizing the maximum value of dispersion of the forecasted value of the studied area.

ORTHOGONAL CENTRAL COMPOSITE EXPERIMENT

With the help of an orthogonal central composite experiment (OCCE), both, the coefficients of regression of second order and the coefficients of higher order can be evaluated. For example,

$$\hat{y} = b_0 + \sum_{i=1}^{m} b_i x_i + \sum_{i<j} b_{ij} x_i x_j + \sum_{i=1}^{m} b_{ii} x_i^2 + \sum_{i<j<k} b_{ijk} x_i x_j x_k + \cdots \quad \text{(II.9.3)}$$

Irrespective of the order of the regression equation, the way of designing does not change.

NUMBER OF TESTS

The number of tests is determined in accordance with formula (II.9.2), as one test ($N_0 = 1$) is conducted in the center of the experiment:

$$N = 2^{m\ p} + 2m + 1. \quad \text{(II.9.4)}$$

DESIGN OF THE EXPERIMENT

For obtaining a diagonal (orthogonal) information matrix, transformation of the variables x_i^2 is necessary under the rule:

$$\tilde{x}_i^2 = x_i^2 - \kappa. \quad \text{(II.9.5)}$$

The variable κ, with the help of which the transformation is done, is determined as

$$\kappa = \frac{2^{m-p} + 2a^2}{2^{m-p} + 2m + 1}, \tag{II.9.6}$$

where a is the size of the star arm determined by the formula

$$a = \pm\sqrt{-2^{m-p-1} + \sqrt{2^{2(m-p-1)} + 2^{m-p-2}(2m+1)}}, \tag{II.9.7}$$

where p is the degree of fractionality (the number of generating correlations) if a fractional factorial experiment is used ($p = 1, 2, 3, \ldots$). In case a full factorial experiment is used, $p = 0$. The design of the experiment for three factors has the form shown in Table II.9.1.

DETERMINATION OF REGRESSION COEFFICIENTS

The zero coefficient is determined in an analogical way as a full factorial experiment:

$$b_0' = \frac{1}{N}\sum_{j=1}^{N} y_j. \tag{II.9.8}$$

Table II.9.1. Orthogonal central composite experiment for three factors

Number	x_0	x_1	x_2	x_3	x_1x_2	x_1x_3	x_2x_3	$\tilde{x}_1^2 = x_1^2 - \kappa$	$\tilde{x}_2^2 = x_2^2 - \kappa$	$\tilde{x}_3^2 = x_3^2 - \kappa$	y
1	+	–	–	–	+	+	+	$1-\kappa$	$1-\kappa$	$1-\kappa$	y_1
2	+	+	–	–	–	–	+	$1-\kappa$	$1-\kappa$	$1-\kappa$	y_2
3	+	–	+	–	–	+	–	$1-\kappa$	$1-\kappa$	$1-\kappa$	y_3
4	+	+	+	–	+	–	–	$1-\kappa$	$1-\kappa$	$1-\kappa$	y_4
5	+	–	–	+	+	–	–	$1-\kappa$	$1-\kappa$	$1-\kappa$	y_5
6	+	+	–	+	–	+	–	$1-\kappa$	$1-\kappa$	$1-\kappa$	y_6
7	+	–	+	+	–	–	+	$1-\kappa$	$1-\kappa$	$1-\kappa$	y_7
8	+	+	+	+	+	+	+	$1-\kappa$	$1-\kappa$	$1-\kappa$	y_8
9	+	a	0	0	0	0	0	$a^2-\kappa$	$-\kappa$	$-\kappa$	y_9
10	+	$-a$	0	0	0	0	0	$a^2-\kappa$	$-\kappa$	$-\kappa$	y_{10}
11	+	0	a	0	0	0	0	$-\kappa$	$a^2-\kappa$	$-\kappa$	y_{11}
12	+	0	$-a$	0	0	0	0	$-\kappa$	$a^2-\kappa$	$-\kappa$	y_{12}
13	+	0	0	a	0	0	0	$-\kappa$	$-\kappa$	$a^2-\kappa$	y_{13}
14	+	0	0	$-a$	0	0	0	$-\kappa$	$-\kappa$	$a^2-\kappa$	y_{14}
15	+	0	0	0	0	0	0	$-\kappa$	$-\kappa$	$-\kappa$	y_{15}

Linear coefficients are determined by the formula:

$$b_i = \frac{\sum_{j=1}^{N} x_{ij} y_j}{2^{m-p} + 2\alpha^2}, \tag{II.9.9}$$

the coefficients before the double interactions are

$$b_{ik} = \frac{\sum_{j=1}^{N} x_{ij} x_{kj} y_j}{2^{m-p}}, \tag{II.9.10}$$

and the coefficients before the transformed variables are

$$b_{ii} = \frac{\sum_{j=1}^{N} \left(x_{ij}^2 - \kappa\right) y_j}{\sum_{j=1}^{N} \left(x_{ij}^2 - \kappa\right)^2}. \tag{II.9.11}$$

DETERMINATION OF REGRESSION COEFFICIENTS' VARIANCES

The variances of regression coefficients are determined based on the variance of reproducibility, S_y^2, determined by an l number of additional tests or, in case of homogenous variances, by the variance of the mean value of y:

$$S^2\{b_o'\} = \frac{1}{N} S_y^2, \tag{II.9.12}$$

$$S^2\{b_i\} = \frac{S_y^2}{2^{m-p} + 2\alpha^2}, \tag{II.9.13}$$

$$S^2\{b_{ik}\} = \frac{S_y^2}{2^{m-p}}, \tag{II.9.14}$$

$$S^2\{b_{ii}\} = \frac{S_y^2}{\sum_{j=1}^{N} \left(x_{ij}^2 - \kappa\right)^2}. \tag{II.9.15}$$

SIGNIFICANCE OF COEFFICIENTS OF REGRESSION EQUATION

Significant are those coefficients, for which the following inequality is valid:

$$|b| > t_T \cdot S\{b\}, \tag{II.9.16}$$

where t_T is the tabular value of Student's t-distribution (determined from Appendix 1) under the following conditions: significance level $a = 0.05$ and degrees of freedom $f = N(n - 1)$ or $(l - 1)$ depending on the way of determination of the reproducibility dispersion,

$S\{b\}$ is standard deviation for the corresponding type of coefficient (zero, before linear members, before interactions or before second orders).

RECORDING OF REGRESSION EQUATION

After elimination of insignificant coefficients, the model has the following general form:

$$y_R = b_o' + \sum_{i=1}^{m} b_i x_i + \sum_{i<j} b_{ij} x_i x_j + \sum_{i=1}^{m} b_{ii}\left(x_i^2 - \kappa\right). \tag{II.9.17}$$

The model gets its final form after conversion from the transformed variables $\tilde{x}_i^2$ to the main variable x_i^2. After opening the brackets, only the zero coefficient changes:

$$b_o = b_o' - \kappa \sum_{i=1}^{m} b_{ii}. \tag{II.9.18}$$

The variance of the zero coefficient b_0 is

$$S^2\{b_o\} = S^2\{b_o'\} + \kappa^2 \sum_{i=1}^{m} S^2\{b_{ii}\}. \tag{II.9.19}$$

VERIFICATION OF ADEQUACY OF THE MODEL

The verification of adequacy is done by the Fisher's criterion as a ratio of the adequacy variance, S_{ad}^2, and the reproducibility variance, S_y^2:

$$F_R = \frac{S_{ad}^2}{S_y^2}. \tag{II.9.20}$$

The adequacy variance is determined as

$$S_{ad}^2 = \frac{\sum_{j=1}^{N}\left(y_{Rj} - y_{Ej}\right)^2}{N - M}, \tag{II.9.21}$$

where y_{Rj} is the value of y determined by the model (expected values),

y_{Ej} is the experimentally obtained values of y,

M is the number of significant coefficients (including the zero coefficient).

The expected value of the Fisher's criterion is compared with the table, one determined under the following conditions F_T ($a = 0.05$, $f_1 = N - M$, $f_2 = N(n - 1)$ or $l - 1$). When $F_R \leq F_T$, the model is adequate.

Example

A mathematical model should be derived describing the influence of the number of abrasion cycles, x_1, and the number of washings, x_2, on the abrasion resistance of knitted fabric.

The main levels of the factors and the intervals of variation are as follows:

- number of abrasion cycles: $x_{0,1} = 1000$, $J_1 = 1000$,
- number of washings: $x_{0,2} = 10$, $J_2 = 10$.

For each of the variants, $n = 9$ tests are conducted.

DETERMINATION OF THE NUMBER OF TESTS

The number of tests is determined according to formula (II.9.4) as the degree of fractionality is $p = 0$:

$$N = 2^2 + 2 \times 2 + 1 = 9.$$

DETERMINATION OF THE SIZE OF THE STAR ARM, α, AND THE VALUE, κ

The size of the star arm is

$$\alpha = \pm\sqrt{-2^{m-p-1} + \sqrt{2^{2(m-p-1)} + 2^{m-p-2}(2m+1)}}$$
$$= \pm\sqrt{-2^{2-1} + \sqrt{2^{2(2-1)} + 2^{2-2}(2 \times 2+1)}} = \pm 1,$$

and the value κ:

$$\kappa = \frac{2^{m-p} + 2\alpha^2}{2^{m-p} + 2m+1} = \frac{2^2 + 2 \times 1^2}{2^2 + 2 \times 2 + 1} = 0.67.$$

PLAN OF THE EXPERIMENT

The plan of the experiment is presented in Table II.9.2.

DETERMINATION OF THE REGRESSION COEFFICIENTS

The regression coefficients are determined according to formulae (II.9.8)–(II.9.11):

- Zero coefficient:

$$b_0' = \frac{1}{9}\sum_{j=1}^{9} y_j = \frac{23070}{9} = 2563;$$

Table II.9.2. Plan of the experiment for the example

Number	x_1	x_2	$\tilde{x}_1^2 = x_1^2 - \kappa$	$\tilde{x}_2^2 = x_2^2 - \kappa$	y_E	S_j^2	$x_1 y$	$x_2 y$	$x_1 x_2 y$	$\tilde{x}_1^2 y$	$\tilde{x}_2^2 y$	y_R	$(y_{Rj} - y_{Ej})^2$
1	–	–	0.33	0.33	4330	78400	–4330	–4330	4330	1428.9	1428.9	4142	35344
2	+	–	0.33	0.33	2330	78400	2330	–2330	–2330	768.9	768.9	2351	441
3	–	+	0.33	0.33	2250	23750	–2250	2250	–2250	742.5	742.5	2173	5929
4	+	+	0.33	0.33	1300	58330	1300	1300	1300	429	429	1432	17424
5	–	0	0.33	–0.67	2850	42450	–2850	0	0	940.5	–1910	3075.5	50850.3
6	+	0	0.33	–0.67	2000	60100	2000	0	0	660	–1340	1809.5	36290.3
7	0	–	–0.67	0.33	3330	78400	0	–3330	0	–2231	1098.9	3445	13225
8	0	+	–0.67	0.33	2080	11070	0	2080	0	–1394	686.4	2001	6241
9	0	0	–0.67	–0.67	2600	24650	0	0	0	–1742	–1742	2641	1681
			–0.03	–0.03	23070	455550	–3800	–4360	1050	–396.9	163.1		167426

(*Continued*)

(Continued)

- Linear coefficients:

$$b_1 = \frac{\sum_{j=1}^{9} x_{1j} y_j}{2^2 + 2\times 1^2} = \frac{-3800}{6} = -633;$$

$$b_2 = \frac{\sum_{j=1}^{9} x_{2j} y_j}{6} = \frac{-4360}{6} = -727;$$

- Coefficients before the double interaction:

$$b_{12} = \frac{\sum_{j=1}^{9} x_{1j} x_{2j} y_j}{2^2} = \frac{1050}{4} = 262.5;$$

- Coefficients before the transformed variables:

$$b_{11} = \frac{\sum_{j=1}^{9} \left(x_{1j}^2 - \kappa\right) y_j}{\sum_{j=1}^{9} \left(x_{1j}^2 - \kappa\right)^2} = \frac{-396.9}{6\times(0.33)^2 + 3\times(-0.67)^2} = -198.5;$$

$$b_{22} = \frac{\sum_{j=1}^{9} \left(x_{2j}^2 - \kappa\right) y_j}{\sum_{j=1}^{9} \left(x_{2j}^2 - \kappa\right)^2} = \frac{163.1}{6\times(0.33)^2 + 3\times(-0.67)^2} = 82.$$

DETERMINATION OF THE VARIANCES OF THE REGRESSION COEFFICIENTS

The variances of the regression coefficients are determined by the variance of reproducibility, S_y^2, which in case of equal variances is the mean value of the variances of the separate tests. The verification of uniformity of the variances is done by the Cochran criterion (from formula (II.6.11)). The expected value is

$$G_R = \frac{S_{\max}^2}{\sum_{j=1}^{k} S_j^2} = \frac{78400}{455550} = 0.172,$$

and the table value is estimated from Appendix 5 under the following conditions $G_T\,(a = 0.05, f_1 = 9 - 1 = 8, f_2 = 9)$. As $G_R = 0.172 \le G_T = 0.277$, the variances are equal and the variance of reproducibility can be determined as the variance of $\overline{y}$:

$$S_y^2 = S_{\bar{y}}^2 = \frac{\sum_{j=1}^{9} S_j^2}{m \cdot N} = \frac{455550}{9 \times 9} = 5624.$$

From this, in accordance with formulae (II.9.12)–(II.9.15), the variances of the separate regression coefficients can be determined:

- Variance of the zero coefficients:

$$S^2\{b_o'\} = \frac{1}{9} S_y^2 = \frac{5624}{9} = 625;$$

- Variance of the linear coefficients:

$$S^2\{b_i\} = \frac{S_y^2}{2^2 + 2 \times 1^2} = \frac{5624}{6} = 937;$$

- Variance of the coefficient before the interaction:

$$S^2\{b_{ik}\} = \frac{S_y^2}{2^2} = \frac{5624}{4} = 1406;$$

- Variance of the coefficients before the second orders of the variables:

$$S^2\{b_{ii}\} = \frac{S_y^2}{\sum_{j=1}^{9} \left(x_{ij}^2 - \kappa\right)^2} = \frac{6524}{2} = 2812.$$

SIGNIFICANCE OF THE COEFFICIENTS OF THE REGRESSION EQUATION

For the separate groups of coefficients, the product $t_T \cdot S\{b\}$ is determined as t_T is determined from Appendix 1 under the following conditions: significance level $a = 0.05$ and degrees of freedom $f = N(n-1) = 9 \times 8 = 72$. The estimated value is $t_T = 2$. The product $t_T \cdot S\{b\}$ is:

- For the zero coefficient:

$$2 \times 25 = 50 < 2563;$$

- For the linear coefficients b_1 and b_2:

$$2 \times 30.6 = 61.2 < |-633|;$$
$$61.2 < |-722|;$$

(*Continued*)

(*Continued*)

- For the coefficients before the interaction b_{12}:

$$2 \times 80 = 160 < |262|;$$

- For the coefficients before the second orders of the variables b_{11} and b_{22}:

$$2 \times 37.5 = 75 < |-198.5|;$$
$$75 < |82|.$$

The verification shows that all coefficients are significant.

RECORD OF THE REGRESSION EQUATION

After the replacement of the obtained regression coefficients, the following mathematical model is obtained:

$$y_R = 2563 - 633x_1 - 727x_2 + 262.5x_1x_2 - 198.5\tilde{x}_1^2 + 82\tilde{x}_2^2.$$

Its final form is obtained after a transition from the transformed variables, $\tilde{x}_i^2$, to the main variables, x_i^2, as a result of which the zero coefficient is changed:

$$b_o = 2563 - 0.67(-198.5 + 82) = 2641,$$

that is, the final record of the model is

$$y_R = 2641 - 633x_1 - 727x_2 + 262.5x_1x_2 - 198.5\tilde{x}_1^2 + 82\tilde{x}_2^2.$$

VERIFICATION OF THE MODEL ADEQUACY

The verification of the adequacy is done by the Fisher criterion as a ratio of the variance of adequacy, S_{ad}^2, and the variance of reproducibility, S_y^2. The expected values, y_{Rj}, are determined in advance according to the final model and they are entered in the penultimate column in Table II.9.2. The squares of the deviations of the expected values from the experimental ones are also determined (the last column in Table II.9.2). With the so determined values, the variance of adequacy is

$$S_{ad}^2 = \frac{\sum_{j=1}^{9}\left(y_{Rj} - y_{Ej}\right)^2}{N - M} = \frac{167426}{9 - 6} = 55809,$$

and the expected value of Fisher's distribution:

$$F_R = \frac{S_{ad}^2}{S_y^2} = \frac{55809}{5624} = 9.92.$$

The tabular value of Fisher's distribution is determined under the following conditions: significance level $a = 0.05$ and degrees of freedom $f_1 = N - M = 3$ and $f_2 = N(n - 1) = 72$. The value estimated from Appendix 3 is $F_T = 2.73$. As $F_R > F_T$, the derived model is not adequate. The reason for this is the big deviations obtained between the measured values and the values calculated according to the model leading to obtaining a high value of the dispersion of adequacy, S_{ad}^2. The approach from here on is a change in the intervals of variation of the variables, a change of the composite plan (e.g., rotatable), or even of the input factors.

ROTATABLE CENTRAL COMPOSITE EXPERIMENT

While the aim of the orthogonal design is to obtain simple formulas for determination of the coefficients of regression, in case of more precise examinations we need a criterion of optimality related to the accuracy of the results obtained. We can judge for the accuracy of the results by the dispersion of the value of the output variable forecasted by the model. The smaller this dispersion is, the more accurate the prognosis is. The points of the rotatable design are located so that when the coordinate axes of the variable are rotated, the distribution of the dispersion does not change. In this way we can ensure the same dispersion at equal intervals from the center of the design.

The composition of a rotatable central composite experiment (RCCE) is analogical to one of the orthogonal central composite experiment—there are points from full and fractional factorial experiments, "star" points, and experiments in the center of the design. The difference is in the method of calculation of the star arm and the number of tests which are chosen so that we have uniform designing.

NUMBER OF TESTS

The total number of tests is

$$N = N_1 + N_2 + N_0, \quad \text{(II.9.22)}$$

where N_1 is the number of tests from the factorial experiment;
N_2 is the number of star points;
N_0 is the number of tests in the center of the experiment.

N_1 and N_2 are determined the same way as in all composite designs and the number of tests in the center of the experiment—with the inequality:

$$\lambda = \frac{m \cdot N}{(m+2)(N - N_o)} \leq 1. \quad \text{(II.9.23)}$$

The number of tests with different number of factors m is given in Table II.9.3.

DESIGN OF THE EXPERIMENT

The size of the star arm is determined as

$$a = 2^{\frac{m-p}{4}}. \tag{II.9.24}$$

Its values for the most commonly used cases are calculated in Table II.9.3. The design of the experiment for two factors has the form shown in Table II.9.4.

Table II.9.3. Number of tests for the rotatable central composite experiment

m	N_1	N_1	N_1	N_1	α	Note
2	4	4	5	13	1.414	
3	8	6	6	20	1.682	
4	16	8	7	31	2.000	
5	32	10	10	52	2.378	
5	16	10	6	32	2.000	Semireplica
6	64	12	15	91	2.828	
6	32	12	9	53	2.378	Semireplica
7	128	14	21	163	3.333	
7	64	14	14	92	2.828	Semireplica
8	128	16	20	164	3.333	Semireplica

Table II.9.4. Design of the rotatable central composite experiment for two factors

Number	x_0	x_1	x_2	x_1x_2	x_1^2	x_2^2	y
1	+	–	–	+	+	+	y_1
2	+	+	–	–	+	+	y_2
3	+	–	+	–	+	+	y_3
4	+	+	+	+	+	+	y_4
5	+	a	0	0	a^2	0	y_5
6	+	$-a$	0	0	a^2	0	y_6
7	+	0	a	0	0	a^2	y_7
8	+	0	$-a$	0	0	a^2	y_8
9	0	0	0	0	0	0	y_9
10	0	0	0	0	0	0	y_{10}
11	0	0	0	0	0	0	y_{11}
12	0	0	0	0	0	0	y_{12}
13	0	0	0	0	0	0	y_{12}

DETERMINATION OF REGRESSION COEFFICIENTS

The determination of regression coefficient by types is done in accordance with the formulas:

$$b_0 = a_1 \sum_{j=1}^{N} y_j - a_2 \sum_{i=1}^{m} \sum_{j=1}^{N} x_{ij}^2 y_j, \tag{II.9.25}$$

$$b_i = a_3 \sum_{j=1}^{N} x_{ij} y_j, \tag{II.9.26}$$

$$b_{ik} = a_4 \sum_{j=1}^{N} x_{ij} x_{kj} y_j, \tag{II.9.27}$$

$$b_{ii} = a_5 \sum_{j=1}^{N} x_{ij}^2 y_j + a_6 \sum_{i=1}^{m} \sum_{j=1}^{N} x_{ij}^2 y_j - a_2 \sum_{j=1}^{N} y_j. \tag{II.9.28}$$

The coefficients from a_1 to a_7 are determined by Table II.9.5.

VARIANCE OF REGRESSION COEFFICIENTS

The first thing to determine is the variance of reproducibility. It is calculated by the tests in the center of the experiment:

$$S_y^2 = \frac{1}{N_o - 1} \sum_{j=1}^{N_o} \left(y_{oj} - \bar{y}_o \right)^2, \tag{II.9.29}$$

Table II.9.5. Values of the auxiliary coefficients

m	a_1	a_2	a_3	a_4	a_5	a_6	a_7	Note
2	0.2000	0.1000	0.1250	0.2500	0.1250	0.01875	0.1437	
3	0.1663	0.0568	0.0732	0.1250	0.0625	0.00688	0.0963	
4	0.1429	0.0357	0.0417	0.0626	0.0313	0.00373	0.0350	
5	0.0988	0.0191	0.0231	0.0313	0.0156	0.00146	0.0171	
5	0.1591	0.0341	0.0417	0.0626	0.0313	0.00284	0.0341	Semireplica
6	0.0625	0.0098	0.0125	0.0156	0.0078	0.00055	0.0084	
6	0.1107	0.0187	0.0231	0.0313	0.0156	0.00122	0.0169	Semireplica
7	0.0398	0.0052	0.0066	0.0077	0.0039	0.00024	0.0041	
7	0.0703	0.0098	0.0125	0.0156	0.0078	0.00049	0.0083	Semireplica
8	0.0450	0.0053	0.0066	0.0077	0.0039	0.00023	0.0041	Semireplica

where $\overline{y}_0$ is the mean value of the tests in the central points:

$$\overline{y}_o = \frac{1}{N_o} \sum_{j=1}^{N_o} y_{oj}. \qquad \text{(II.9.30)}$$

From the variance of reproducibility, the variances of the coefficients are also determined:

$$S^2\{b_o\} = a_1 S_y^2 \qquad \text{(II.9.31)}$$

$$S^2\{b_i\} = a_3 S_y^2 \qquad \text{(II.9.32)}$$

$$S^2\{b_{ik}\} = a_4 S_y^2 \qquad \text{(II.9.33)}$$

$$S^2\{b_{ii}\} = a_7 S_y^2 \qquad \text{(II.9.34)}$$

SIGNIFICANCE OF COEFFICIENTS OF REGRESSION EQUATION

Significant are those regression coefficients b, for which $|b| > t_T \cdot S\{b\}$, where t_T ($a = 0.05$, $f = N_o - 1$) is determined from Appendix 1 under the specified conditions.

VERIFICATION OF ADEQUACY OF THE MODEL

The adequacy dispersion is determined as

$$S_{ad}^2 = \frac{Q_1 - Q_2}{N - M - (N_o - 1)}, \qquad \text{(II.9.35)}$$

where Q_1 and Q_2 are sums of the squares of the deviations, respectively of the calculated y_{Rj} from the measured y_{Ej} values and of the values y_{0j} measured in the center from their mean value $\overline{y}_o$. They are determined in the following way:

$$Q_1 = \sum_{j=1}^{N} \left(y_{Rj} - y_{Ej}\right)^2, \qquad \text{(II.9.36)}$$

$$Q_2 = \sum_{j=1}^{N_o} \left(y_{oj} - \overline{y}_o\right)^2. \qquad \text{(II.9.37)}$$

The calculated value of Fisher's criterion is

$$F_R = \frac{S_{ad}^2}{S_y^2}, \qquad \text{(II.9.38)}$$

and the tabular value with which we compare F_T ($a = 0.05$, $f_1 = N - M - (N_o - 1)$, $f_2 = N_o - 1$). With $F_R \leq F_T$, the model is adequate.

Example

A mathematical model should be derived describing the influence of pressure P (kPa), temperature T (°C), and quantity of steam, determined according to the position of the adjusting bolt H (mm), on time for moisture–thermal processing.

The main levels of the factors and the variance intervals are as follows:

- pressure: $x_{0,1} = 470$ kPa, $J_1 = 130$ kPa
- temperature: $x_{0,2} = 143$°C, $J_2 = 13$°C
- position of the adjusting bolt: $x_{0,3} = 7$ mm, $J_3 = 3$ mm.

A rotatable central composite experiment is used for derivation of the model. The number of tests and the size of the star arm are determined according to Table II.9.2. The levels of separate factors in natural and coded values are shown in Table II.9.6.

Table II.9.6. Levels of factors in natural and coded values

	x_1		x_2		x_3	
	$J_1 = 130$		$J_2 = 13$		$J_3 = 3$	
Levels	**Coded**	**Natural**	**Coded**	**Natural**	**Coded**	**Natural**
$-\alpha$	−1.682	251	−1.682	121	−1.682	2
−1	−1	340	−1	130	−1	4
0	0	470	0	143	0	7
1	1	600	1	156	1	10
α	1.682	689	1.682	165	1.682	12

The design of the experiment, the measured values of the output parameter, as well as the supporting products for determination of regression coefficients are given in columns 1–14 of Table II.9.7.

DETERMINATION OF REGRESSION COEFFICIENTS

Regression coefficients are determined by formulae (II.9.25)–(II.9.28):

$$b_0 = a_1 \sum_{j=1}^{20} y_j - a_2 \sum_{i=1}^{3} \sum_{j=1}^{20} x_{ij}^2 y_j = 0.1663 \times 471.5 - 0.0568(361.26 + 358.43 + 299.02) = 20.55,$$

$$b_1 = a_3 \sum_{j=1}^{20} x_{1j} y_j = 0.0732 \times (-91.0) = -6.66,$$

$$b_2 = a_3 \sum_{j=1}^{20} x_{2j} y_j = 0.0732 \times (-89.69) = -6.57,$$

(Continued)

(*Continued*)

Table II.9.7. Design of the experiment, measured values of the output parameter and supporting products for determination of regression coefficients

Number	x_1	x_2	x_3	x_1y	x_2y	x_3y	x_1x_2y	x_1x_3y	x_2x_3y	x_1^2y	x_2^2y	x_3^2y	y_E	y_R	$(y_{Rj}-y_{Ej})^2$	$(y_{oj}-\bar{y}_o)^2$
1	2	3	4	5	6	7	8	9	10	11	12	13	14	15	16	17
1	–1	–1	–1	–33.5	–33.5	–33.5	33.5	33.5	33.5	33.5	33.5	33.5	33.5	33.86	0.127	
2	1	–1	–1	18.5	–18.5	–18.5	–18.5	–18.5	18.5	18.5	18.5	18.5	18.5	18.78	0.081	
3	–1	1	–1	–20.5	20.5	–20.5	–20.5	20.5	–20.5	20.5	20.5	20.5	20.5	18.98	2.320	
4	1	1	–1	7.5	7.5	–7.5	7.5	–7.5	–7.5	7.5	7.5	7.5	7.5	7.40	0.009	
5	–1	–1	1	–44.5	–44.5	44.5	44.5	–44.5	–44.5	44.5	44.5	44.5	44.5	44.15	0.126	
6	1	–1	1	29	–29	29	–29	29	–29	29	29	29	29	29.07	0.005	
7	–1	1	1	–28.5	28.5	28.5	–28.5	–28.5	28.5	28.5	28.5	28.5	28.5	29.27	0.586	
8	1	1	1	18	18	18	18	18	18v	18	18	18	18	17.69	0.095	
9	1.682	0	0	29.44	0	0	0	0	0	49.51	0	0	17.5	17.28	0.047	
10	–1.682	0	0	–66.44	0	0	0	0	0	111.75	0	0	39.5	39.69	0.037	
11	0	1.682	0	0	27.75	0	0	0	0	0	46.68	0	16.5	16.94	0.198	
12	0	–1.682	0	0	–66.44	0	0	0	0	0	111.75	0	39.5	39.03	0.222	
13	0	0	1.682	0	0	44.57	0	0	0	0	0	74.97	26.5	26.13	0.134	
14	0	0	–1.682	0	0	–14.30	0	0	0	0	0	24.05	8.5	8.83	0.108	
15	0	0	0	0	0	0	0	0	0	0	0	0	21.5	20.55	0.907	0.840
16	0	0	0	0	0	0	0	0	0	0	0	0	20.5	20.55	0.002	0.007
17	0	0	0	0	0	0	0	0	0	0	0	0	20	20.55	0.300	0.340
18	0	0	0	0	0	0	0	0	0	0	0	0	20	20.55	0.300	0.340
19	0	0	0	0	0	0	0	0	0	0	0	0	21	20.55	0.205	0.174
20	0	0	0	0	0	0	0	0	0	0	0	0	20.5	20.55	0.002	0.007
				–91.0	–89.69	70.28	7	2	–3	361.26	358.43	299.02	471.5		5.810	1.708

$$b_3 = a_3 \sum_{j=1}^{20} x_{3j} y_j = 0.0732 \times 70.28 = 5.14,$$

$$b_{12} = a_4 \sum_{j=1}^{20} x_{1j} x_{1j} y_j = 0.125 \times 7 = 0.875,$$

$$b_{13} = a_4 \sum_{j=1}^{20} x_{1j} x_{3j} y_j = 0.125 \times 2 = 0.25,$$

$$b_{23} = a_4 \sum_{j=1}^{20} x_{2j} x_{3j} y_j = 0.125 \times (-3) = -0.375,$$

$$b_{11} = a_5 \sum_{j=1}^{20} x_{1j}^2 y_j + a_6 \sum_{i=1}^{3} \sum_{j=1}^{20} x_{1j}^2 y_j - a_2 \sum_{j=1}^{20} y_j$$
$$= 0.0625 \times 361.26 + 0.00688 \times (361.26 + 358.43 + 299.02) - 0{,}0568 \times 471.5 = 2.81,$$

$$b_{22} = a_5 \sum_{j=1}^{20} x_{2j}^2 y_j + a_6 \sum_{i=1}^{3} \sum_{j=1}^{20} x_{2j}^2 y_j - a_2 \sum_{j=1}^{20} y_j$$
$$= 0.0625 \times 358.43 + 0.00688 \times (361.26 + 358.43 + 299.02) - 0.0568 \times 471.5 = 2.63,$$

$$b_{33} = a_5 \sum_{j=1}^{20} x_{3j}^2 y_j + a_6 \sum_{i=1}^{3} \sum_{j=1}^{20} x_{3j}^2 y_j - a_2 \sum_{j=1}^{20} y_j$$
$$= 0.0625 \times 299.02 + 0.00688 \times (361.26 + 358.43 + 299.02) - 0.0568 \times 471.5 = -1.08.$$

DETERMINATION OF VARIANCE OF REPRODUCIBILITY

It is calculated from the tests in the center of the experiment by formula (II.9.29) as the mean value $\bar{y}_o = 20.58$ is determined in advance:

$$S_y^2 = \frac{1}{6-1} \sum_{j=1}^{6} \left(y_{oj} - 20.58 \right)^2 = 0.342.$$

DETERMINATION OF VARIANCES OF REGRESSION COEFFICIENTS

For the different groups of coefficients, the dispersions of the regression coefficients are:

- For the zero coefficient:

$$S^2\{b_o\} = 0.1663 \times 0.342 = 0.0568,$$

(*Continued*)

(*Continued*)

- For the linear coefficients:

$$S^2\{b_i\} = 0.0732 \times 0.342 = 0.025,$$

- For the coefficients before the interactions:

$$S^2\{b_{ik}\} = 0.125 \times 0.342 = 0.0427,$$

- For the coefficients before the second orders of the variables:

$$S^2\{b_{ii}\} = 0.0693 \times 0.342 = 0.0237.$$

VERIFICATION OF SIGNIFICANCE OF REGRESSION COEFFICIENTS

The value of Student's t-distribution is determined from Appendix 1 with significance level $a = 0.05$ and degrees of freedom $f = 6 - 1 = 5$ as the accounted value is $t_T = 2.57$. The products $t_T \cdot S\{b\}$ are determined and compared with the calculated coefficients:

- For the zero coefficient:

$$2.57 \times 0.238 = 0.613 < 20.55,$$

- For the linear coefficients b_1, b_2, and b_3:

$$2.57 \times 0.158 = 0.406 < |-6.66|,$$

$$0.406 < |-6.565|,$$

$$0.406 < |5.14|,$$

- For the coefficients before the interactions b_{12}, b_{13}, and b_{23}:

$$2.57 \times 0.207 = 0.531 < |0.875|,$$

$$0.531 > |0.25|,$$

$$0.531\ |-0.385|,$$

- For the coefficients before the second orders of the variables b_{11}, b_{22}, and b_{33}:

$$2.57 \times 0.154 = 0.396 < |2.81|,$$

$$0.396 < |2.63|,$$

$$0.396 < |-1.08|.$$

Insignificant are only coefficients b_{13} and b_{23} the absolute value of which is smaller than the critical one.

RECORDING OF THE MODEL

After elimination of the insignificant factors, the model has the following form:

$$y = 20.55 - 6.66x_1 - 6.565x_2 + 5.14x_3 + 0.875x_1x_2 + 2.81x_1^2 + 2.63x_2^2 + 1.08x_3^2.$$

VERIFICATION OF MODEL ADEQUACY

Determination of the sum Q_1 (sum of column 16):

$$Q_1 = \sum_{j=1}^{20} \left(y_{Rj} - y_{Ej} \right)^2 = 5.81.$$

Determination of the sum Q_2 (sum of column 17):

$$Q_2 = \sum_{j=1}^{6} \left(y_{oj} - \bar{y}_o \right)^2 = 1.708.$$

Determination of the adequacy variance S_{ad}^2:

$$S_{ad}^2 = \frac{Q_1 - Q_2}{N - M - (N_o - 1)} = \frac{5.81 - 1.708}{20 - 8 - (6 - 1)} = 0.586.$$

The expected Fisher's criterion is

$$F_R = \frac{S_{ad}^2}{S_y^2} = \frac{0.586}{0.342} = 1.715.$$

The table value of Fisher's distribution is $F_T(a = 0.05, f_1 = 20 - 8 - (6 - 1) = 7, f_2 = 6 - 1 = 5) = 4.876$. As $F_R = 1.715 < 4.876 = F_T$, the model is adequate.

OPTIMAL COMPOSITE EXPERIMENT

The structure of optimal composite designs ensures maximal values of the determinant of the information matrix with a size of the star arm $\alpha = \pm 1$. It is also established that, except for the cases with two factors, designs are most effective if no tests that are made in the center of the plan.

NUMBER OF TESTS

Considering the specificity of the composition of optimal composite designs, the number of tests can be recorded as

$$N = N_1 + N_2 + N_o, \tag{II.9.39}$$

as $N_o = 1$ (with $n \le 2$) and $N_o = 0$ (with $n > 2$).

DESIGN OF THE EXPERIMENT

The design of the experiment with three factors is shown in Table II.9.8.

DETERMINATION OF REGRESSION COEFFICIENTS

The regression coefficients are determined by the formulae below:

$$b_o = a\sum_{j=1}^{N} y_j + p\sum_{i=1}^{m}\sum_{j=1}^{N} x_{ij}^2 y_j, \tag{II.9.40}$$

$$b_i = e\sum_{j=1}^{N} x_{ij} y_j, \tag{II.9.41}$$

$$b_{ik} = g\sum_{j=1}^{N} x_{ij} x_{kj} y_j, \tag{II.9.42}$$

$$b_{ii} = c\sum_{j=1}^{N} x_{ij}^2 y_j + d\sum_{\substack{i=1\\ j\neq i}}^{m}\sum_{j=1}^{N} x_{ij}^2 y_j + p\sum_{j=1}^{N} y_j. \tag{II.9.43}$$

The values of coefficients *a*, *p*, *c*, *d*, *e*, and *g* are given in Table II.9.9.

Table II.9.8. Design of the optimal composite experiment for three factors

Number	x_1	x_2	x_3	y
1	–	–	–	y_1
2	+	–	–	y_2
3	–	+	–	y_3
4	+	+	–	y_4
5	–	–	+	y_5
6	+	–	+	y_6
7	–	+	+	y_7
8	+	+	+	y_8
9	+	0	0	y_9
10	–	0	0	y_{10}
11	0	+	0	y_{11}
12	0	–	0	y_{12}
13	0	0	+	y_{13}
14	0	0	–	y_{14}

Table II.9.9. Values of the auxiliary coefficients

m	N	a	p	c	d	e	g	Note
2	9	0.555555	–0.333333	0.500000	0.000000	0.166666	0.250000	
3	14	0.406250	–0.156250	0.406250	–0.093750	0.100000	0.125000	
4	24	0.229165	–0.062500	0.395833	–0.104166	0.055555	0.062500	
5	42	0.158203	–0.033203	0.408203	–0.091797	0.029412	0.031250	
5	26	0.160156	–0.035156	0.410156	–0.089844	0.055555	0.062500	Semireplica
6	76	0.120625	–0.020625	0.420625	–0.079375	0.015151	0.015625	
6	44	0.121250	–0.021250	0.421250	–0.078750	0.029412	0.031250	Semireplica
7	78	0.097656	–0.014329	0.430989	–0.069010	0.015151	0.015625	Semireplica

VARIANCE OF REGRESSION COEFFICIENTS

The variance of reproducibility, S_y^2, is usually determined by l additional tests in the center of the experiment, and from it the variances of the coefficients are determined:

$$S^2\{b_o\} = a\,S_y^2, \tag{II.9.44}$$

$$S^2\{b_i\} = e\,S_y^2, \tag{II.9.45}$$

$$S^2\{b_{ik}\} = g\,S_y^2, \tag{II.9.46}$$

$$S^2\{b_{ii}\} = c\,S_y^2. \tag{II.9.47}$$

SIGNIFICANCE OF COEFFICIENTS OF REGRESSION EQUATION

Significant are those coefficients for which $|b| > t_T \cdot S\{b\}$, where $t_T\,(a = 0.05, f = l - 1)$.

VERIFICATION OF ADEQUACY

The adequacy verification is done in the same way as in the orthogonal experiment.

Example
The influence of the distance in the main drawing field L (mm), the drawing i in it and the force on the pressing rollers (daN) in a drawing frame on the index of irregularity I of the cotton slivers is examined. A mathematical model describing their influence should be derived. Two measurements are made for each of the variants as the mean values are shown in Table II.9.10.

(*Continued*)

(*Continued*)

Table II.9.10. Experimental data for the example

	16 daN			26.65 daN			37.3 daN		
I	**36 mm**	**39 mm**	**42 mm**	**36 mm**	**39 mm**	**42 mm**	**36 mm**	**39 mm**	**42 mm**
$i = 4.80$	8.41	7.84	8.87	8.22	7.66	8.53	8.22	7.90	8.78
$i = 5.68$	8.20	7.92	8.50	7.47	7.73	8.37	7.95	7.77	8.84
$i = 6.54$	8.62	7.55	9.06	8.07	7.44	8.68	9.08	7.73	8.51

Two additional tests (a total of four) are conducted in the center of the experiment, as a result of which the standard deviation of the examined value $S_y = 0.1$ is determined. An optimal composite plan is constructed and presented in Table II.9.11.

DETERMINATION OF THE REGRESSION COEFFICIENTS

The regression coefficients are determined according to formulae (II.9.40)–(II.9.43).

$$b_o = a\sum_{j=1}^{14} y_j + p\sum_{i=1}^{3}\sum_{j=1}^{14} x_{ij}^2 y_j = 0.40625\times 116.18 - 0.15625\times(85.39+84.65+85.24) = 7.311,$$

$$b_1 = e\sum_{j=1}^{14} x_{1j}y_j = 0.1\times 1.79 = 0.179,$$

$$b_2 = e\sum_{j=1}^{14} x_{2j}y_j = 0.1\times 0.77 = 0.077,$$

$$b_3 = e\sum_{j=1}^{14} x_{3j}y_j = 0.1\times(-0.52) = -0.052,$$

$$b_{12} = g\sum_{j=1}^{14} x_{1j}x_{2j}y_j = 0.125\times(-1.15) = -0.144,$$

$$b_{13} = g\sum_{j=1}^{14} x_{1j}x_{3j}y_j = 0.125\times(-0.91) = -0.114,$$

$$b_{23} = g\sum_{j=1}^{14} x_{2j}x_{3j}y_j = 0.125\times 0.09 = 0.024,$$

$$b_{11} = c\sum_{j=1}^{14} x_{1j}^2 y_j + d\sum_{\substack{i=1\\ j\neq i}}^{3}\sum_{j=1}^{14} x_{ij}^2 y_j + p\sum_{j=1}^{14} y_j$$

$$= 0.40625\times 85.39 - 0.09375\times(84.65+85.24) - 0.15625\times 116.18 = 0.609,$$

Table II.9.11. Optimal composite plan for the example

Number	x_1	x_2	x_3	y_E	x_1y	x_2y	x_3y	x_1x_2y	x_1x_3y	x_2x_3y	x_1^2y	x_2^2y	x_3^2y	y_R	$\left(y_{oj}-\bar{y}_o\right)^2$
1	–1	–1	–1	8.41	–8.41	–8.41	–8.41	8.41	8.41	8.41	8.41	8.41	8.41	8.26	0.023
2	1	–1	–1	8.87	8.87	–8.87	–8.87	–8.87	–8.87	8.87	8.87	8.87	8.87	9.13	0.068
3	–1	1	–1	8.62	–8.62	8.62	–8.62	–8.62	8.62	–8.62	8.62	8.62	8.62	8.54	0.006
4	1	1	–1	9.06	9.06	9.06	–9.06	9.06	–9.06	–9.06	9.06	9.06	9.06	8.84	0.047
5	–1	–1	1	8.22	–8.22	–8.22	8.22	8.22	–8.22	–8.22	8.22	8.22	8.22	8.48	0.070
6	1	–1	1	8.78	8.78	–8.78	8.78	–8.78	8.78	–8.78	8.78	8.78	8.78	8.90	0.015
7	–1	1	1	9.08	–9.08	9.08	9.08	–9.08	–9.08	9.08	9.08	9.08	9.08	8.77	0.095
8	1	1	1	8.51	8.51	8.51	8.51	8.51	8.51	8.51	8.51	8.51	8.51	8.62	0.011
9	1	0	0	8.37	8.37	0	0	0	0	0	8.37	0	0	8.10	0.073
10	–1	0	0	7.47	–7.47	0	0	0	0	0	7.47	0	0	7.74	0.073
11	0	1	0	7.44	0	7.44	0	0	0	0	0	7.44	0	7.55	0.012
12	0	–1	0	7.66	0	–7.66	0	0	0	0	0	7.66	0	7.55	0.012
13	0	0	1	7.77	0	0	7.77	0	0	0	0	0	7.77	7.85	0.006
14	0	0	–1	7.92	0	0	–7.92	0	0	0	0	0	7.92	7.85	0.006
				116.18	1.79	0.77	–0.52	–1.15	–0.91	0.19	85.39	84.65	85.24		0.517

(Continued)

(*Continued*)

$$b_{22} = c\sum_{j=1}^{14} x_{2j}^2 y_j + d\sum_{\substack{i=1\\ j\neq i}}^{3}\sum_{j=1}^{14} x_{ij}^2 y_j + p\sum_{j=1}^{14} y_j$$

$$= 0.40625 \times 84.65 - 0.09375 \times (85.39 + 85.24) - 0.15625 \times 116.18 = 0.239,$$

$$b_{33} = c\sum_{j=1}^{14} x_{3j}^2 y_j + d\sum_{\substack{i=1\\ j\neq i}}^{3}\sum_{j=1}^{14} x_{ij}^2 y_j + p\sum_{j=1}^{14} y_j$$

$$= 0.40625 \times 85.24 - 0.09375 \times (85.39 + 84.65) - 0.15625 \times 116.18 = 0.534.$$

DETERMINATION OF THE VARIANCES OF THE REGRESSION COEFFICIENTS

The variance of reproducibility is: $S_y^2 = 0.1^2 = 0.01$, from which, in accordance with the formulae (II.9.44)–(II.9.47), the variances of the coefficients are determined:

- for the zero coefficient:

$$S^2\{b_o\} = a\,S_y^2 = 0.40625 \times 0.01 = 0.004;$$

- for the linear coefficients:

$$S^2\{b_i\} = e\,S_y^2 = 0.1 \times 0.01 = 0.001;$$

- for the coefficient before the interaction:

$$S^2\{b_{ik}\} = g\,S_y^2 = 0.125 \times 0.01 = 0.001;$$

- for the coefficients before the second orders of the variables:

$$S^2\{b_{ii}\} = c\,S_y^2 = 0.\,40625 \times 0.01 = 0.004.$$

VERIFICATION OF THE SIGNIFICANCE OF THE COEFFICIENTS OF THE REGRESSION EQUATION

From Appendix 1, the value of the Student's distribution is determined with a significance level $a = 0.05$ and degrees of freedom $f = 4 - 1 = 3$, as the estimated values is $t_T = 3.182$. The products $t_T \cdot S\{b\}$ are determined and compared to the calculated coefficients:

- For the zero coefficient:

$$3.182 \times 0.0632 = 0.2011 < 7.311;$$

- For the linear coefficients b_1, b_2, and b_3:

$$3.182 \times 0.0316 = 0.1005 < |0.179|;$$

$$0.1005 > |0.077|;$$

$$0.1005 > |-0.052|;$$

- For the coefficients before the interactions b_{12}, b_{13}, and b_{23}:

$$3.182 \times 0.0316 = 0.1005 < |-0.144|;$$

$$0.1005 < |-0.114|;$$

$$0.1005 > |0.024|;$$

- For the coefficients before the second orders of the variables b_{11}, b_{22}, and b_{33}:

$$3.182 \times 0.0632 = 0.2011 < |0.609|;$$

$$0.2011 < |0.239|;$$

$$0.2011 < |-0.534|.$$

The only insignificant coefficients are b_2, b_3, and b_{23}, the absolute value of which is lower than the critical one.

RECORD OF THE MODEL

After the elimination of the insignificant coefficients the model gets the following form:

$$y = 7.311 + 0.179x_1 - 0.144x_1x_2 - 0.144x_1x_3 + 0.609x_1^2 + 0.239x_2^2 + 0.534x_3^2.$$

VERIFICATION OF THE MODEL ADEQUACY

The verification of the adequacy is done by the Fisher criterion, according to formula (II.9.21):

$$S_{ad}^2 = \frac{\sum_{j=1}^{N} \left(y_{Rj} - y_{Ej} \right)^2}{N - M} = \frac{0.517}{14 - 7} = 0.074,$$

and the calculated value of Fisher's criterion is:

$$F_R = \frac{S_{ad}^2}{S_y^2} = \frac{0.074}{0.01} = 7.39.$$

The tabular value of Fisher's distribution is determined from Appendix 3 under the following conditions F_T ($a = 0.05$, $f_1 = N - M = 14 - 7 = 7$, $f_2 = l - 1 = 4 - 1 = 3$) = 8.89. As $F_R < F_T$, the model is adequate.

II.10. Optimization of Target Function

The description of optimum area is done with the aim of studying the behavior of the target function around the extremum and to find an optimal working regime of the examined system. Different approaches exist for solving the optimization task. The most commonly used methods of optimization are:

1. *Classical method of function analysis*—the minimum or the maximum of the target function is determined by differentiation and equalization of its first derivative to zero. This method is appropriate for polynomial regression models of second order as a system of linear equations which are easy to solve are obtained after the differentiation.
2. *Canonical analysis*—the equation is reduced to a canonical form through the methods of linear algebra. The reduction to a canonical form is an introduction of new axes concurring with the axes of the surface described by the equation of regression. Depending on the signs of coefficients of the canonical equation, the type of the surface of response can be determined. The method is commonly used because it is notable for its demonstrability.
3. *Method of Lagrange multipliers*—it is used when the target functions are more than one. For one of the functions, we search for an optimum and constraints are placed for the others. The constraints may only be exact values.
4. *Methods of linear programming*—for target functions which are linear with respect to the studied factors. The constraints can also be linear functions.
5. *Methods of nonlinear programming*—they are applied when the target function and the constraints are nonlinear functions of the factors. They are divided into three main groups:
   - gradient methods;
   - gradientless methods; and
   - methods of random searching.
6. *Dynamic programming*—used for optimization of multistage processes. The mathematical description of each stage can be obtained with the help of experimental statistical methods.

CANONICAL ANALYSIS

In reduction of the model to a canonical form, the linear members and the interactions are removed. The canonical form of the equation is

$$y_R - y_{RS} = B_{11}x_{k1}^2 + B_{22}x_{k2}^2, \tag{II.10.1}$$

where y_R is the value of the output parameter before the transformation,
y_{RS} is the extremal value of the output parameter,
B_{11} and B_{22} are the new coefficients before x_1^2 and x_2^2.

The new axes are obtained in two stages: first, through translation of the extremal point S (the axes x_1' and x_2') and second, through rotation around point S of the angle α until their concurrence with the axes of the surface (Figure II.10.1). The axes x_1'' and x_2'' are the axes of the canonical equation and for this reason they are written as x_{k1} and x_{k2}.

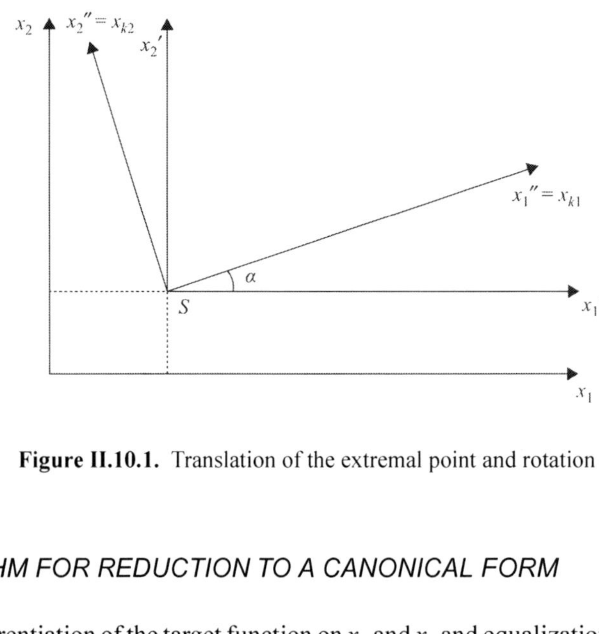

Figure II.10.1. Translation of the extremal point and rotation of the axes.

ALGORITHM FOR REDUCTION TO A CANONICAL FORM

1. Differentiation of the target function on x_1 and x_2 and equalization of the partial derivative to zero.
2. Solving the system of linear equations and determination of the values x_{1S} and x_{2S} with which an extremum is obtained. Substitution of the obtained values of x_{1S} and x_{2S} and determination of the extremal value y_{RS}. Determination of the rotation angle α:

$$\alpha = 0.5\text{arctg}\frac{b_{12}}{b_{11} - b_{22}}. \tag{II.10.2}$$

3. Determination of the regression coefficients in the canonical equation B_{11} and B_{22}. Two methods exist for determination of B_{11} and B_{22}:

$$B^2 - (b_{11} + b_{22})B + (b_{11}b_{22} - 0.25b_{12}^2) = 0. \tag{II.10.3}$$

The regression coefficients in the canonical equation are:

$$B_{11} = \frac{P}{2} - \sqrt{\left(\frac{P}{2}\right)^2 - Q} \tag{II.10.4}$$

and

$$B_{22} = \frac{P}{2} + \sqrt{\left(\frac{P}{2}\right)^2 - Q}, \tag{II.10.5}$$

as:

$$P = b_{11} + b_{22}, \tag{II.10.6}$$

and

$$Q = b_{11}b_{22} - 0.25b_{12}^2. \tag{II.10.7}$$

Method II—Trigonometric

The regression coefficients in the canonical equation are determined as

$$B_{11} = b_{11}\cos^2 \alpha + b_{12}\sin\alpha\cos\alpha + b_{22}\sin^2 \alpha; \tag{II.10.8}$$

$$B_{22} = b_{11}\sin^2 \alpha - b_{12}\sin\alpha\cos\alpha + b_{22}\cos^2 \alpha. \tag{II.10.9}$$

The verification of the correctness of calculations is

$$\sum b_{ii} = \sum B_{ii}. \tag{II.10.10}$$

DETERMINATION OF THE SURFACE TYPE

The surface type is determined by the signs of B_{11} and B_{22}. The following cases are possible:

1. B_{11} and B_{22} are with the same sign: In this case the surface is an elliptic paraboloid and the equipotential lines are ellipses with the equation:

$$\frac{x_1^2}{a^2} + \frac{x_2^2}{b^2} = 1. \tag{II.10.11}$$

If B_{11} and $B_{22} > 0$, the extremum is minimum; if B_{11} and $B_{22} < 0$, the extremum is maximum; if $|B_{22}| > |B_{11}|$, the main axis of the ellipses is x_{k1}, and vice versa.

2. B_{11} and B_{22} are with different signs: In this case the surface is a hyperbolic paraboloid and the equipotential lines are hyperbolas. If $|B_{22}| > |B_{11}|$, the main axis of the hyperbolas is on x_{k1}; if $B_{11} < 0$ and $B_{22} > 0$, the target function y decreases from the center to x_{k1} and increases to x_{k2}; if $B_{11} > 0$ and $B_{22} < 0$, the target function y increases from the center to x_{k1} and decreases to x_{k2}.
3. One of the coefficients is zero: The surface is a fixed elevation and the contour lines are parallel lines. If $B_{11} = 0$, the lines are parallel to x_{k1}; if $B_{22} = 0$, the lines are parallel to x_{k2}.
4. One of the coefficients is zero, and the center of the coordinate system is infinity: Then we obtain an ascending elevation and the contour lines are parabolas.

In Figure II.10.2 an elliptic paraboloid is shown and in Figure II.10.3 the equipotential lines of this surface are shown.

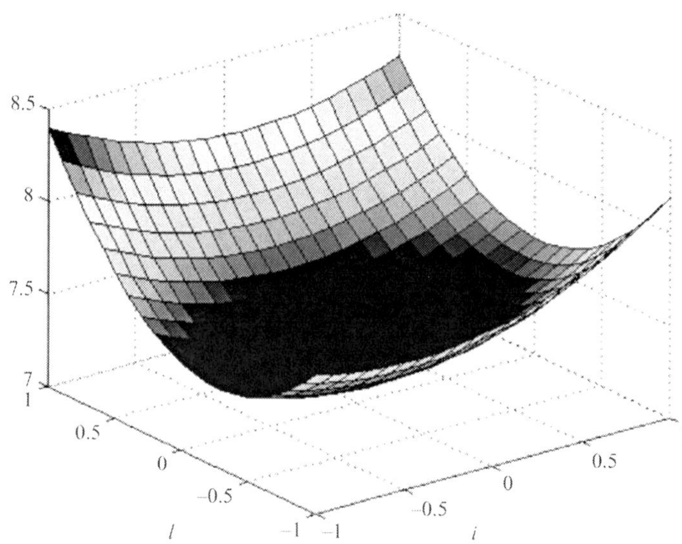

Figure II.10.2. Elliptic paraboloid surface.

In Figure II.10.4 a hyperbolic paraboloid is shown and in Figure II.10.5 the corresponding equipotential lines are shown.

It can be seen that in both cases the extremum is minimum. The equipotential lines being the contour of the surface sections with planes parallel to the plane Ox_1x_2 to distances from it chosen in advance, allow for the location of the minimum to be determined visually. In the first case it is located at the center of the ellipse corresponding to $y = 7.4$ and in the second case at a level of $x_1 = -1$ and x_2 in the middle point of the segment line obtained at the crossing of the parabola $y = 7.6$ with the axis x_2.

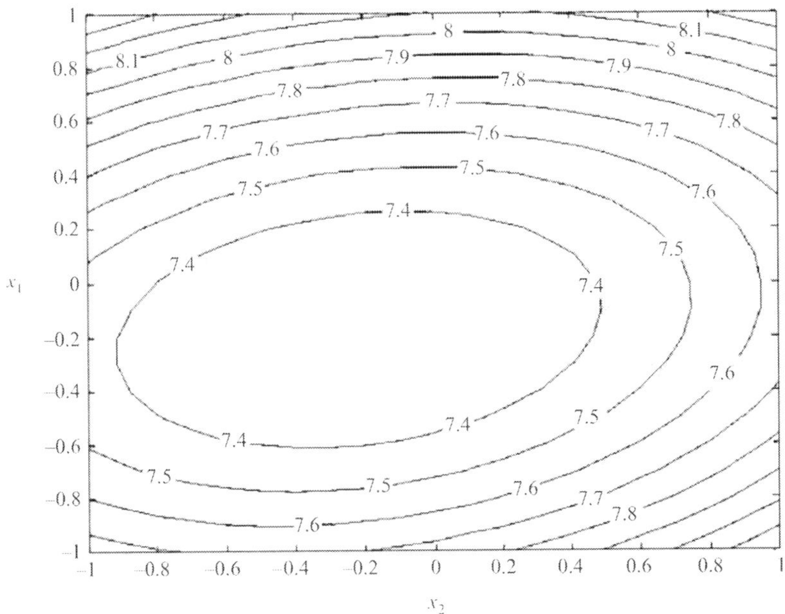

Figure II.10.3. Equipotential lines of elliptic paraboloid surface.

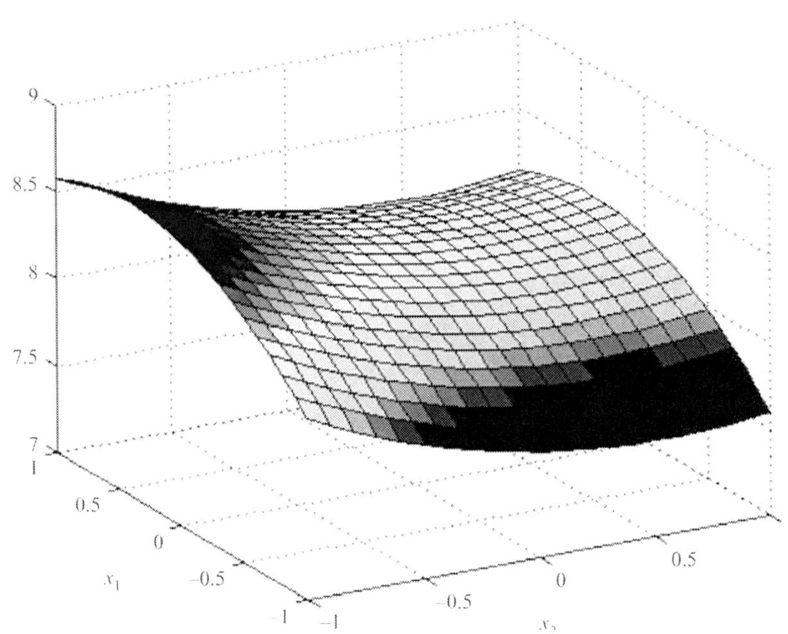

Figure II.10.4. Hyperbolic paraboloid surface.

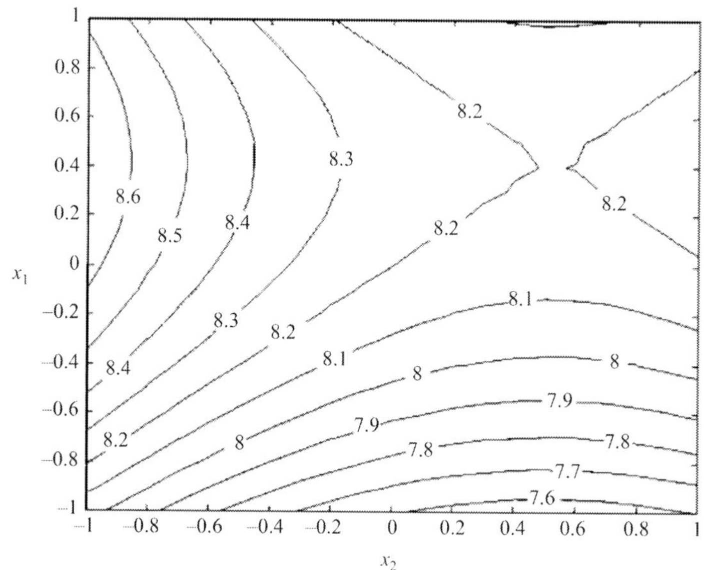

Figure II.10.5. Equipotential lines of hyperbolic paraboloid surface.

Example

The following mathematical model describing the influence of the distance x_1 (mm) and the drawing rate x_2 in a cylinder drawing device on the irregularity CV (%) of the cotton sliver at the exit of the cylinder drawing frame is derived with the help of optimal composite design:

$$y = 7.311 + 0.181x_1 + 0.078x_2 - 0.143x_1x_2 + 0.608x_1^2 + 0.238x_2^2.$$

The optimal technological setting should be found as: $x_{01} = 39$ mm, $J_1 = 3$ mm, $x_{02} = 5.68$ mm, and $J_2 = 0.87$.

DIFFERENTIATION OF TARGET FUNCTION

$$\frac{\partial y}{\partial x_1} = 0.181 - 0.143x_{2S} + 2 \times 0.608x_{1S}$$

$$\frac{\partial y}{\partial x_2} = 0.078 - 0.143x_{1S} + 2 \times 0.238x_{2S}$$

(Continued)

SOLVING THE SYSTEM OF LINEAR EQUATIONS

$$\left| \begin{array}{l} 0.181 - 0.143x_{2S} + 2 \times 0.608x_{1S} = 0 \\ 0.078 - 0.143x_{1S} + 2 \times 0.238x_{2S} = 0 \end{array} \right.$$

$$x_{1S} = -0.174 \qquad x_{2S} = -0.216$$

DETERMINATION OF THE EXTREME VALUE OF OUTPUT PARAMETER

$$y_S = 7.311 + 0.181(-0.174) + 0.078(-0.216) - 0.143(-0.174)(-0.216)$$
$$= 0.608(-0.174)^2 + 0.238(0.216)^2 = 7.287$$

DETERMINATION OF THE ROTATION ANGLE

$$\alpha = 0.5\text{arctg}\frac{0.143}{0.608 - 0.238} = 10.6^\circ$$

DETERMINATION OF REGRESSION COEFFICIENTS IN CANONICAL EQUATION

By Method I

$$P = b_{11} + b_{22} = 0.608 + 0.238 = 0.846$$

$$Q = b_{11}b_{22} - 0.25b_{12}^2 = 0.608 \times 0.238 - 0.25 \times 0.143^2 = 0.1396$$

$$B_{11} = \frac{0.846}{2} + \sqrt{\left(\frac{0.846}{2}\right)^2 - 0.1396} = 0.621$$

$$B_{22} = \frac{0.846}{2} - \sqrt{\left(\frac{0.846}{2}\right)^2 - 0.1396} = 0.225$$

By Method II

$$B_{11} = 0.608\cos^2(10.6^\circ) + 0.143\sin(10.6^\circ)\cos(10.6^\circ) + 0.238\sin^2(10.6^\circ) = 0.621$$
$$B_{22} = 0.608\sin^2(10.6^\circ) - 0.143\sin(10.6^\circ)\cos(10.6^\circ) + 0.238\cos^2(10.6^\circ) = 0.225$$

(*Continued*)

(Continued)

ADDITIONAL VERIFICATION OF CALCULATION CORRECTNESS

$$0.608 + 0.238 = 0.621 + 0.225 = 0.846$$

DETERMINATION OF SURFACE TYPE

As B_{11} and B_{22} are with the same sign, the surface is an elliptic paraboloid (Figure II.10.6).

The surface drawing is done as the values of y are calculated with the variation of x_1 and x_2 within the diapason [–1; 1] with the model:

$$y = 7.311 + 0.181x_1 + 0.078x_2 - 0.143x_1x_2 + 0.608x_1^2 + 0.238x_2^2.$$

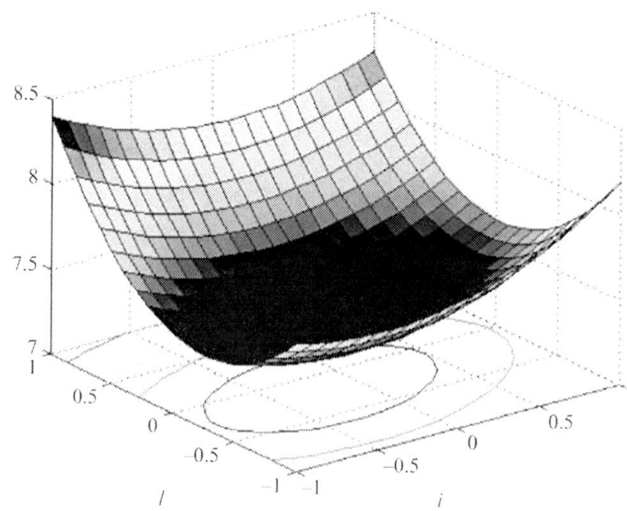

Figure II.10.6. Surface described by the model.

The equipotential lines are drawn using the canonical form of equation (II.10.1) in which the obtained values y_{RS} and the coefficients B_{11} and B_{22} are replaced:

$$y_R - 7.287 = 0.621x_{k1}^2 + 0.225x_{k2}^2.$$

Both sides of the equation are divided by (y_R – 7.287) for an equation of an ellipse to be obtained. We get:

$$1 = \frac{x_{k1}^2}{a^2} + \frac{x_{k2}^2}{b^2},$$

(Continued)

where:

$$a^2 = \frac{y_R - 7.287}{0.621} \quad \text{and} \quad b^2 = \frac{y_R - 7.287}{0.225}$$

and a and b are the values of the semiaxes of the ellipse determined for the different values of y_R. The values of a and b in the variation y_R are shown in Table II.10.1.

Table II.10.1. Values of the semiaxes in the variation of y_R

y_R	$y_R - y_{RS}$	a^2	b^2	a	b
7.4	0.113	0.182	0.502	0.427	0.709
7.5	0.213	0.343	0.947	0.586	0.973
7.6	0.313	0.504	1.391	0.710	1.179
7.7	0.413	0.665	1.836	0.816	1.355
…	…	…	…	…	…

DETERMINATION OF THE OPTIMAL PARAMETERS OF THE CYLINDER DRAWING DEVICE

From the calculations made, and as it is visible from the graphics, the lower irregularity $CV = 7.287\%$ is obtained with levels of the factors in coded values $x_{1S} = -0.174$ and $x_{2S} = -0.216$. In natural values:

- distance $x_{1S,nat} = x_{01} + x_{1S}J_1 = 39 + (-0.174) \times 39 = 38.5$ mm
- drawing rate $x_{2S,nat} = x_{02} + x_{2S}J_2 = 5.68 + (-0.216) \times 0.87 = 5.38.$

II.11. Taguchi Methods

Taguchi methods are statistical methods for improvement of the quality of manufactured articles. They can be grouped in three main directions:

- designed experiments
- off-line methods of control
- loss function.

INNOVATIONS IN THE SPHERE OF THE DESIGNED EXPERIMENT

Taguchi creates a technique for quality improvement using methods of experiment design and combines them with statistical analysis. This approach allows for the quality requirements to be embedded in the stage of design and be followed in every stage of the article's life cycle. He suggests the following steps in the conducting of experimental investigations:

1. *Definition of the problem*—defining the problem to be solved and the task that will be solved with the experiment.
2. *Determination of the target*—choosing the output characteristics that will be investigated and optimized, as well as the method of measurement.
3. *Conducting of intense mental investigation*—determining the manageable and unmanageable factors, defining the investigation area, and the levels of the factors. In the beginning, during the conducting of sifting experiments, the use of as many factors as possible is recommended at the expense of the interactions.
4. *Design of the experiment*—setting the values of the control factors and their interactions and of the nuisance factors. Taguchi divides the factors into manageable (control) and nuisance factors, as he includes the nuisance factors in the designed experiment in the outer arrays.
5. *Conducting of the experiment*—conducting of the tests and experimental data collection.
6. *Information analysis*—calculating the characteristic indices [target measure (TM) and nuisance measure (NM)] and analyzing with the use of appropriate methods.
7. *Result interpretation*—identifying the managing factors of deviation and the target managing factors. Choosing of optimal levels. For the managing factors of deviation, the optimal levels are the ones maximizing the nuisance measure (minimizing the dispersion in the model reaction), and for the target managing factors the ones bringing the

mean values of the reaction close to the target value. Foreseeing the characteristics of the system at optimal values of the factors.

8. *Conducting of a confirmation experiment*—confirming that the newly chosen parameters improve the characteristic of the article through an experiment.

If the expected results are not confirmed or they are not satisfactory, additional experiments are conducted repeating steps 3–8.

OFF-LINE METHODS OF CONTROL

According to Taguchi, the best way to eliminate deviations and inconsistencies is by including the quality requirements at the time of the product design and the production process. The strategy he developed includes three stages:

- *System design:* The article's functional characteristics, the choice of materials, technologies, and production devices are determined. It is done on the basis of an extensive analysis of market demands and production capabilities. As a result, the product conception is created.
- *Parameter design:* On the grounds of the conception, the parameters of the product are designed. The optimal parameters of the articles and the processes are determined with which the influence of disturbing factors is minimal. The process of giving the system stability and insensitiveness to random disturbances is called robustification.
- *Parameter limits (tolerance) design:* After designing the system and the parameters, the influence of different parameters on the functioning of the system can be evaluated. The attention is focused on the decrease and control of the deviations in narrow limits.

SYSTEM DESIGN

A number of methods are used for system design based on creativity and innovation. Creative methods can be divided into four groups:

1. *Intuitive methods (logical–emotional):* A part of them are the different variants of brainstorming (classical brainstorming, Method 635, and reverse brainstorming), synectics (analogies method), and so on.
2. *Morphological methods (system–combinative):* The optimal combination of all known attributes and characteristics of the analyzed object should be found with them. The most important characteristics are determined first and arranged in morphological matrices. Different combinations of parameters according to the separate criteria are analyzed and the best option should be found.
3. *Algorithmic (system–programming) methods:* They are based on the objective laws of technical systems and psychological regularity of creative activity (Mueller's systematic heuristics, Polovinkin's generalized heuristic algorithm, and so on).
4. *Integral methods (system–complex):* Methods combining the best of the above listed methods.

The methods of Quality Function Deployment (QFD) and Failure Mode and Effect Analysis (FMEA) are also widely used.

Quality Function Deployment

QFD is a powerful and often-used method for granting the quality of design, production, and realization of articles. The method was first advanced in Japan by Professor Oshiumi in 1966. In 1972, Professor Akao applied the approach in practice in the Mitsubishi–Kobe shipyard and for this reason he is considered to be the initiator of the method. QFD is defined as a "system for transformation of user demands into characteristics appropriate for the relative company at each stage of the investigation from product design and development through production, distribution, planning, sales marketing, and maintenance." At first, the method had been applied only at the stage of articles design where it had proven to be very effective for decrease in the time for development and in problems causing frequent project revisions. Due to its efficiency and universality, it has begun to be used at the next stages of the articles' life cycle as well. The stages in the complex application of the method are:

1. *Designing of the article:* Transformation of specific consumer requirements to engineering value characteristics with the relevant dimensions.

   Example: The requirement of the consumer for a strong fabric is transformed to warp strength A (N) and weft strength B (N).
2. *Detailed design:* Transformation of engineering characteristics to characteristics of details, tangles, and elements of the article.

   Example: The warp strength A (N) is determined by the strength of the warp yarn C (cN/tex), the warp density D (number of threads/10 cm), the weave E, and so on.
3. *Process design:* Transformation of the detailed characteristics to characteristics of the technological process.

   Example: The strength of the warp threads C (cN/tex) is determined by the speed of the spinning machine, the twists of the yarn, the centering, and so on.
4. *Design of production:* Combination of technological and production characteristics.

   Example: Composition of spinning plan for production of warp threads, technological and machine parameters in the preparation of the threads for weaving, in the process of weaving, in the finishing processing, and so on.

The main characteristics of the method are:

- It is based on the consumer requirements.
- The results of the previous stage are requirements for the next.
- Participation of specialists from different areas and on different levels (horizontal and vertical) as decisions are made with consensus.

The main mean of the method is the "House of quality" (Figure II.11.1). Three- to five-degree index-line is used most often for ranging of relations between the elements. In composition of the matrices, the following order of tasks is applied:

1. *Consumer requirements*—consumer wishes are identified by preparation of a full list of the requirements determined on the grounds of preliminary data from polls and other researches. Both, wanted and unwanted characteristics are recorded. The collected information is summarized and classified into a few levels: a small number of essential requirements, worked out in details on a second and, if necessary, on a third level.
   1.1. *Level of importance of requirements*—consumer wishes are arranged according to their importance and marked with a number.
2. *Characteristics of quality*—this is the stage of "deployment of the characteristics of quality" in which the requirements are transformed to measurable characteristics. The characteristics are also presented on levels and numbered.
   2.1. *Direction of improvement of characteristics*—an arrow is used to show whether improvement of the quality is obtained with an increase (↑) or a decrease (↓) in the value of the parameter or with a nominal value (0).
   2.2. *Autocorrelation matrix of characteristics*—it shows the correlations between the separate characteristic (positive or negative).
3. *Relations between requirements and characteristics*—with the so-called matrix of quality, the level of conformity of the article's characteristics with the consumer requirements is shown.
4. *Competitive power*—the matrix shows the consumer evaluation of the article according to the different requirements compared to the ones of the competition. In case of lack of information, data from service activity can be used.
5. *Values and dimensions*—the target values and the dimensions of the different characteristics or their features are specified.
   5.1. *Level of technical importance*—the importance of the separate characteristics and their features is given in absolute and relative values. The absolute values can be obtained also as the levels of dependency from matrix 3 are multiplied by 9 for a strong relation, by 4 for a medium one, by 1 for a weak one and are summed by columns.
   5.2. *Level of technical difficulty*—the difficulty in improvement of the quality of the respective feature is evaluated by a conditional index line.

Depending on the case, additional matrices can be used, for example for control, expenses or other technical and economic features.

Failure Mode and Effect Analysis

FMEA is an analytical engineering method used for investigation and elimination of potential problems, faults, and errors. In its nature, it is a preventive method for granting the quality of articles and processes. It was developed in the early 1960s by NASA for the needs of the American space program. Afterward, it has been applied in aircraft and automobile industry and now it has a realization in all branches of industry. In some countries the method is standardized and obligatory for branches such as atomic power engineering. The essence of the method is the detailed investigation of the article or the process. Any possible error, fault, or nonconformity is analyzed and documented, their influence on the quality, the risk of their occurrence, and the

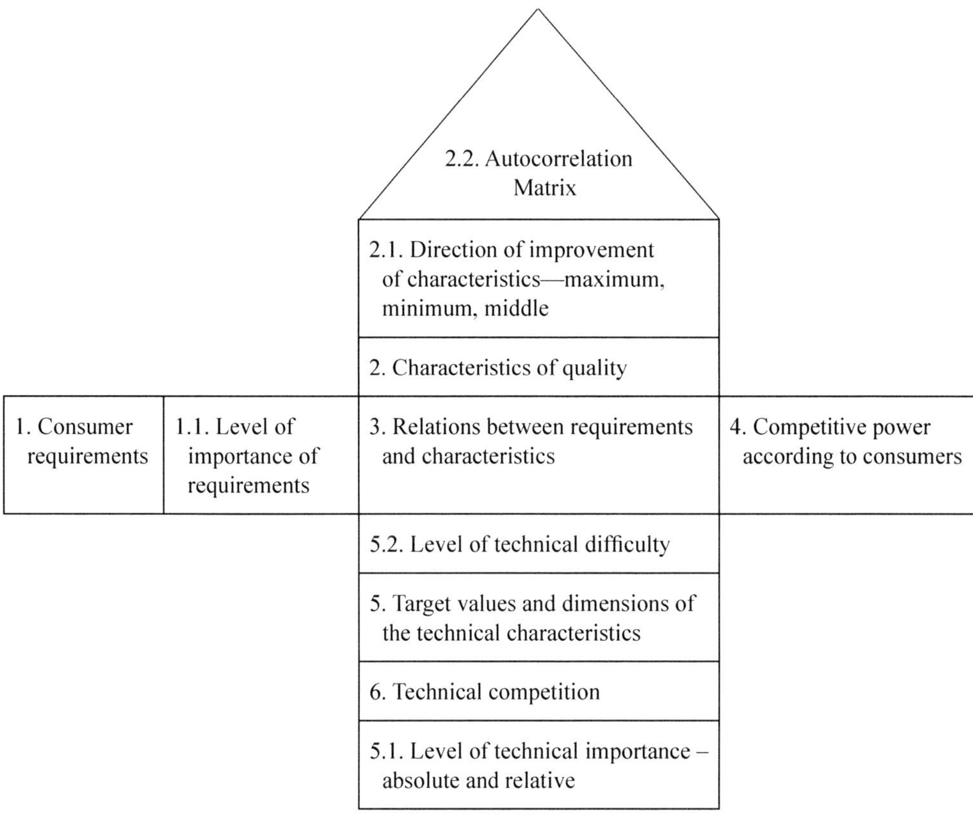

Figure II.11.1. The "House of quality."

possibilities for their location and prevention are evaluated. With a simple scheme for categorization, the conditional value of the risk is determined for every potential reason and it recognizes the possibility for recurrence, the effect, and the possible consequences. The possibility to find the fault before the article reaches the consumer is determined. The analysis is done for each element of the article, or each stage of the process, as special attention is paid to the ones on which safety, security, and the other major characteristics depend. The development and the introduction of the method are done by a specialized group in which constructors, technologists, and representatives of the quality control department participate. It is especially important that specialists with considerable experience are included in the group.

Design FMEA is used for the analysis of potential error-related construction and materials, for example:

- impossibility for manufacturing or assembly;
- low reliability;
- faults in case of regular or irregular exploitation; and
- use of inappropriate materials.

Process FMEA is applied in product development for analysis of possible errors in production. In the analysis, the full technical and technological documentation, the prospects of the machines, devices, and equipment are used. Each stage of the process is discussed by the order of the technological operations and the following are determined:

- errors and faults due to machines, equipment, and technology;
- errors in control and testing; and
- errors in packing and dispatching.

The main results are written down in a FMEA chart (Figure II.11.2).

FMEA chart				
For construction			**For process**	
FMEA group	Working out	Development	Production	Quality
<u>Confirmed</u>			Date..................	
Signature................				
Detail name..................			Draft	
Number..................				
Producer/supplier..................				
Changes: Number.................. from.................. year				

Analysis of situation							**Estimation**			
№	**Detail**	**Function**	**Defect**	**Consequences**	**Reasons**	**Control**	**OCC**	**SEV**	**DET**	**RPN**
1	2	3	4	5	6	7	8	9	10	11

Actions for improvement			**Estimation**			
Recommended action	**Executor**	**Realized actions**	**OCC**	**SEV**	**DET**	**RPN**
12	13	14	15	16	17	18

Figure II.11.2. FMEA chart.

Its' development is done in two stages:

1. Condition analysis (columns 1–11)

   Columns 1 to 7 show, respectively, the number, name, function of the detail (or the technological operation), possible faults or errors, supposed consequences, potential reasons, and the provided control. In the following four columns, an evaluation of the condition is made by three differentiated (columns 8, 9, and 10) and one complex index (column 11).

   - *OCC* is the probability of occurrence of a fault or an error in the process. It is determined on the basis of project calculations, results of testing, and experience from previous developments. It is evaluated with a point system. An evaluation 1 means low probability and 10 means very high probability.
   - *SEV* shows the level of influence of the fault. The level of the possible consequences is evaluated as, with an evaluation 1 the fault does not have an influence on the quality of the article and with 10 it has a diverse display and severe consequences.
   - *DET* is the probability of detection of the fault (the error) before the article reaches the customer or the next operation of the process. Evaluation 1 means the fault will be found with certainty and 10 means it will reach the customer.
   - *RPN* is the coefficient of risk. It is obtained as a product of the three measures above. Depending on the value obtained, it can be determined for which of the faults and the errors urgent measures have to be taken (with $RPN > 125$) and which ones do not affect the (with $RPN < 40$) process.

2. Actions for improvement

   In columns 12–14, the measures that have to be taken regarding the high-risk faults and errors, the executors of the recommended activities, and the results of their realization are shown. In columns 15–18, a new evaluation of the condition after the measures taken is done. Activities are undertaken in case of low values of risk and also in case of changes in the construction, the technology, the methods, and the means of control, and so on. The chart is disseminated in the departments of the company for notes and recommendations. After finalization of the discussion and recording of the changes, the chart is approved and sent for execution to the relevant departments and persons.

   The method is characteristic for its high efficiency and universality which makes it an important element of the quality management systems.

PARAMETER DESIGN

The aim of parameter design is to improve production process and construct the product through easily controllable factors. These factors have to be given in such a way that the deviation of the product characteristics from the mean value is minimized and the mean value itself takes the required target value. Through setting of optimal levels of these factors, the product, to a large extent, becomes independent from the variation of the transitory conditions. With parameter design one can "eliminate the unfavorable effect of the cause instead of the cause of the unfavorable effect." As a result of this, high-quality articles with stable features are obtained. Moreover, if the method is applied systematically in the preproduction stage, it can considerably decrease the prolonged testing for determination of the most cost effective production conditions thus saving money and materials.

Division of Factors

According to Taguchi, the behavior of a product is characterized by two types of factors (parameters):

1. *Control (manageable) parameters*—the ones the values of which can easily be changed by the constructor or the technologist.
2. *Noncontrol (nuisance) parameters*—the ones that are the sources of deviations which are most often a result of production, organization, and management conditions. In the ideal case, the characteristics of the product should be insensitive to the variation of the nuisance parameters.

The control factors are divided in turn into:

1. *Target control factors (signal factors)*—such factors that affect the mean values of the characteristics.
2. *Deviation control factors*—such factors that affect the deviation of the characteristics.
3. *Price factors*—such factors that affect neither the mean values nor the deviations but can be set as economic requirements.

A scheme of the factors by groups is given in Figure II.11.3.

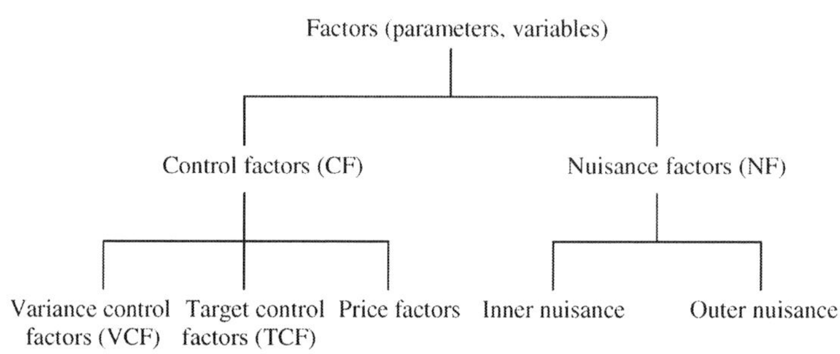

Figure II.11.3. Classification of the acting factors.

What distinguishes the approach of Taguchi from traditional methods of control of tolerances and quality management is exactly the focus on dispersion. The idea is for the dispersion to be decreased through a change of the factors that control it and for the required mean levels to be maintained through setting of target control factors.

Elimination or Reduction of the Effect of the Nuisance Factors

In this stage, the aim is for the effect of the nuisance factors to be reduced or eliminated. The influence of the nuisance factors is simulated during the experiment through their systematic variation for each level of the investigated control factors. It is considered that the nuisance factors can be controlled within the experiment.

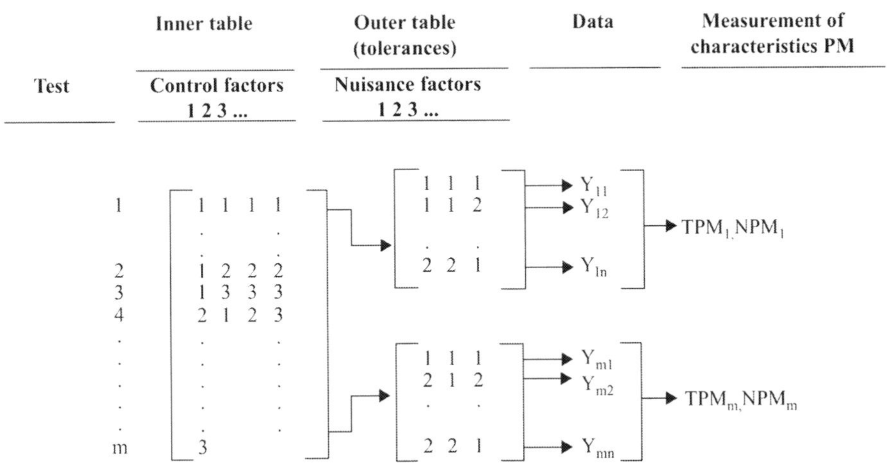

Figure II.11.4. Matrix form of the experiment design.

The control parameters vary in accordance with the table of the designed experiment. Usually a partial orthogonal table is used in which each value of a given parameter is combined with each value of all the other parameters (factors) equal number of times. For each row (a combination of control factors), observations are made when there is a variation of the nuisance factors. The orthogonal table (matrix) is used for determination of the levels and the combinations of the nuisance factors. Thus the experiment simulates the effect of the nuisance factors on each value of the control parameters as the value in which the dispersion is minimized is determined. The general form of the experiment design can be given in the form of a diagram (Figure II.11.4).

The inner matrix (of the control parameters) has m rows and the outer matrix (of the nuisance factors) has n rows and the number of observations of the value of the output variable is $m{\cdot}n$. They are used for calculation of measures characterizing each of the m inner rows.

The inner matrix is selected and not a random subset in the area of the nuisance factors. The experimental levels of the nuisance parameters are chosen in such a way that the whole area is covered. If the distribution of the nuisance factor, N_i, has a mean value, m_i, a standard deviation, S_i, and a symmetrical distribution, Taguchi recommends:

- with a postulation that N_i has a linear effect, the factor has to be varied in the following two levels: $(m_i - S_i)$ and $(m_i S_i)$
- with a postulation that N_i has a nonlinear effect, the factor has to be varied in the following three levels: $\left(m_i - S_i\sqrt{3/2}\right)$, m_i, and $\left(m_i + S_i\sqrt{3/2}\right)$.

The experiment can be conducted physically or by computer simulations provided that the reaction of the model can be numerically calculated. When a physical experiment is conducted, in many of the cases it is impossible or too expensive for it to be conducted in compliance with the design of setting of the parameters shown in Figure II.11.4. Then the conducting of a partial experiment is recommended as the parameters are varied on two or no more than three levels. If the nuisance factors cannot be controlled, Taguchi recommends that as many random observations as possible should be made for each row as much as the economic preconditions allow it.

This approach is also used in the classical experiment designing. The difference is that in this case the repetition of every test is used for determination of the dispersion index while in the classical case the repetitions are used for determination of the total experimental error.

Characteristic Measures

From the test results for each combination of control factors, two characteristic indices are calculated:

1. *Target performance measure* (TPM), which shows the mean value of the characteristic for each group of values. It is used to determine the control factors affecting the mean value of the output parameter, as well as for its adjustment to the desired target value. It is determined by the mean arithmetic value for each test.
2. *Nuisance performance measure* (NPM), which shows the variation in the studied characteristics. It serves to determine the control parameters and find the optimal combination of values of the parameters in which the dispersion (or the effect of the nuisance factors) is minimized.

For determination of the nuisance measure in the presence of one target value, Taguchi recommends the use of the so-called signal-to-noise ratio which is calculated from the reciprocal value of the variation coefficient. For each of the experimental tests, the *SNR* measure is

$$SNR = 10\log_{10}\left(\frac{\overline{x}}{S}\right)^2, \qquad \text{(II.11.1)}$$

where $\overline{x}$ is the mean value of the target measure,
S is the standard deviation of the target measure.

When the desired characteristics of the reaction of the system are "the lesser the better" (e.g., in case of faulty articles), Taguchi recommends the use of the measure:

$$\eta = -10\log_{10}\left\{\frac{1}{n}\sum x_i^2\right\}, \qquad \text{(II.11.2)}$$

where n is the number of tests,
x_i is the values of the separate tests.

When the requirements are "the more the better," Taguchi recommends the determination of the measure:

$$\theta = -10\log_{10}\left\{\frac{1}{n}\sum x_i^{-2}\right\}. \qquad \text{(II.11.3)}$$

When the characteristics are measured by a binary index-line (fit/unfit, success/failure), what is suggested is the determination of:

$$z = 10\log_{10}\left\{\frac{p}{1-p}\right\}, \qquad \text{(II.11.4)}$$

where p is the share of the desired (favorable) cases. The measures recommended by Taguchi have to be used carefully. If there is no functional correlation between the mean value and the dispersion, even only the logarithm of the dispersion S^2 can be used for evaluation of the noise. It is appropriate for the *SNR* measure to be applied in the presence of linear correlation between the mean value and the dispersion. By correlating the two values, we can achieve independency of the deviation control factors from the target control factors.

Interactions between the Factors

Taguchi states that "a man who does not *believe* in the existence of *nonlinear effects* is a man removed from reality." However, he considers that engineers can determine such levels of the studied factors that some (or all) effects of interaction can become insignificant for the specific experiment. In this sense, the partial design is not only more economical but it can also prove to be more appropriate for the respective experiment. At the same time Taguchi strongly recommends the conducting of a confirmation experiment after the determination of the optimal values.

In Table II.11.1 we can see the orthogonal table that can be considered as three main factors (A, B, and C) and all possible interactions between them or seven main factors (A, B, C, D, E, F, and G) without considering the interactions.

The processing of the test results is done in the following order:

1. The mean dispersion for each test is determined:

$$S_j^2 = \frac{1}{n}\sum_{i=1}^{n} x_{ij}^2. \tag{II.11.5}$$

2. The logarithms of the mean dispersion are taken and as a result of that there is a transformation from linear scale to decibels:

$$z_j = -10\log_{10}\left(S_j^2\right) \text{ [dB]}. \tag{II.11.6}$$

Table II.11.1. Taguchi's orthogonal table

Test	A	B	C	D $A \times B$	E $A \times C$	F $B \times C$	G $A \times B \times C$	Results
1	1	1	1	1	1	1	1	$x_{11} \ldots x_{1n}$
2	1	1	2	1	2	2	2	$x_{21} \ldots x_{2n}$
3	1	2	1	2	1	2	2	...
4	1	2	2	2	2	1	1	...
5	2	1	1	2	2	1	2	...
6	2	1	2	2	1	2	1	...
7	2	2	1	1	2	2	1	...
8	2	2	2	1	1	1	2	$x_{81} \ldots x_{8n}$

3. Separation of the desired signal from the noise. The mean value of z_j is calculated and then also the differences dz_j:

$$\overline{z} = \frac{1}{8}\sum_{j=1}^{8} z_j; \tag{II.11.7}$$

$$dz_j = z_j - \overline{z}. \tag{II.11.8}$$

4. After obtaining the values of the desired signal dz_j, the processing continues in the classical method of conducting of dispersion analysis: the total sum of the squares, the sums of the squares of the separate factors, and the evaluations of the dispersions are determined.
5. Following the procedure, the dispersion analysis is finalized with determination of the residual dispersion and verification of hypotheses with the Fisher's criterion. In order to avoid the significant number of experiments, Taguchi suggests that the sums of the separate factors are compared and the ones having smaller values, that is, have a nonessential influence, are eliminated. The eliminated factors are ranked with the residual dispersion (the experiment error) and the verification with the Fisher's criterion is done only for the remaining essential factors. With a postulation that the factors *E*, *F*, and *G* were eliminated as nonessential, a dispersion table is constructed (Table II.11.2).
6. Determination of the relative share of each factor in the total dispersion:

$$C_k = \frac{(Q_k - S_R^2)}{Q} 100. \tag{II.11.9}$$

where *k* is the respective factor.
The expression in the brackets is the net variation of the factor.

Table II.11.2. Taguchi's dispersion table

Source of dispersion	Degrees of freedom	Sum of the squares	Dispersion evaluation	F_R
Factor *A*	$f_A = 1$	Q_A	$S_A^2 = Q_A$	$\frac{S_A^2}{S_R^2}$
Factor *B*	$f_B = 1$	Q_B	$S_B^2 = Q_B$	$\frac{S_B^2}{S_R^2}$
Factor *A* × *B*	$f_{AB} = 1$	Q_{AB}	$S_{AB}^2 = Q_{AB}$	$\frac{S_{AB}^2}{S_R^2}$
Factor *D*	$f_D = 1$	Q_D	$S_D^2 = Q_D$	$\frac{S_D^2}{S_R^2}$
Residual errors	$f_R = 3$	Q_R	$S_R^2 = \frac{Q_R}{3}$	
Total	7	Q		

7. Determination of the optimal level for each factor. When there is a combination in the group of the essential factors, the determination of the optimal levels starts from it. The mean values of the desired signal are calculated for all combinations of levels, for the example this is the combination $A \times B$:

$$S(A_1B_1) = \frac{\delta z_1 + \delta z_2}{2}, \tag{II.11.11}$$

$$S(A_1B_2) = \frac{\delta z_3 + \delta z_4}{2}, \tag{II.11.13}$$

$$S(A_2B_1) = \frac{\delta z_5 + \delta z_6}{2}, \tag{II.11.12}$$

$$S(A_2B_2) = \frac{\delta z_7 + \delta z_8}{2}. \tag{II.11.13}$$

The optimal levels of each of the factors A and B correspond to the sum with the maximum value. If this is the second sum, the optimal level for the factor A is 1 and for B is 2. The mean values of the desired signal for the single factors are determined in an analogical way, and from the maximum values the optimal level:

$$S(k_1) = \frac{\sum \delta z_{j1}}{4}, \tag{II.11.14}$$

$$S(k_2) = \frac{\sum \delta z_{j2}}{4}. \tag{II.11.15}$$

8. Determination of the quality improvement. For this purpose, the mean value of the differences δz_j is calculated:

$$\overline{\delta z} = \frac{1}{8}\sum_{j=1}^{8} \delta z_j \tag{II.11.16}$$

and the quality index QI_k is determined for each factor k:

$$QI_k = Q_{k,\max} - \overline{\delta z}. \tag{II.11.17}$$

The value of QI_k shows the quality improvement due to the optimally determined level of the respective factor. If the value δz corresponding to the combination of levels before the optimization is known (usually this is the first row of Table II.11.1), the general quality improvement can also be determined:

$$QI = \sum Q_{k,\max} - \delta z_1. \tag{II.11.18}$$

The obtained value for QI is in decibels. The transition to a linear scale is done through the reverse transformation:

$$QI_L = 10^{\frac{-Q}{10}}. \qquad \text{(II.11.19)}$$

If the variance S^2 of the variable x_{ij} before the optimization is known, the value of the variance after the optimization can be determined with the help of QI_L:

$$S_{opt}^2 = QI_L \cdot S^2. \qquad \text{(II.11.20)}$$

The results of the optimization in finance units can be shown in the same way. With loss of the consumer L_0 before the optimization, they will be:

$$L = QI_L \cdot L_0, \qquad \text{(II.11.21)}$$

and the difference $(L - L_0)$ will be the profit of the quality improvement.

For optimization of the costs, the loss function is used and it will be presented later.

DESIGN OF THE PARAMETER TOLERANCES (TOLERANCE PLAN)

If it is not possible to achieve elimination of the effect from the nuisance (internal and external) factors through parameter design, Taguchi recommends design of the tolerance of the article's parameters. The design is done by the procedures described above but additional factors are also included that may have been eliminated in the initial stages due to the high price, or the difficulty for conducting of the necessary experiments. If the effect of the nuisance factors is not eliminated even with this approach, what can be suggested is an investigation of the tolerance of the products' components. The optimal levels of the factors determined during the previous stage are kept but the tolerance of factors essential to the price and the quality is decreased. Thus the deviation of the reaction is decreased to admissible levels. The decrease in the tolerance of the essential factors allows a certain extension of the tolerance for components that are not so responsible.

Tolerance determination is a systematic procedure that identifies the most important nuisance factors and suggests a price-effective way of their control. This is the stage in which we make decisions for elimination of those nuisance factors, the effect of which cannot be eliminated. Tolerance design is done after the stage of parameter design and only under the condition that the previous procedures did not give the desired effect.

LOSS FUNCTION

Taguchi defines quality as "losses to society after the moment of manufacturing of a given product." These losses are a result of a consumers' dissatisfaction and have a direct display as losses from reclamations and so on, as well as an indirect one from reputation loss. According to Taguchi, a product causes loss not only when it is not in conformity with the requirements but also when it deviates from its target value. He considers that the purpose of each program for quality improvement should be the minimization of variations in the product's characteristics with respect to the target value. The higher the deviation is from this value, the bigger the losses for consumers and producers are.

The Taguchi loss function *L*(*X*) is a square correlation presenting a relation between financial loss and deviation of the product's functional characteristics (Figure II.11.5):

$$L(X) = \frac{M(X-t)^2}{D^2}, \tag{II.11.22}$$

where *M* is the loss of the consumer in money in case of exceeding the consumer tolerance,
X is characteristic of the product,
t is target value,
D is consumer tolerance.

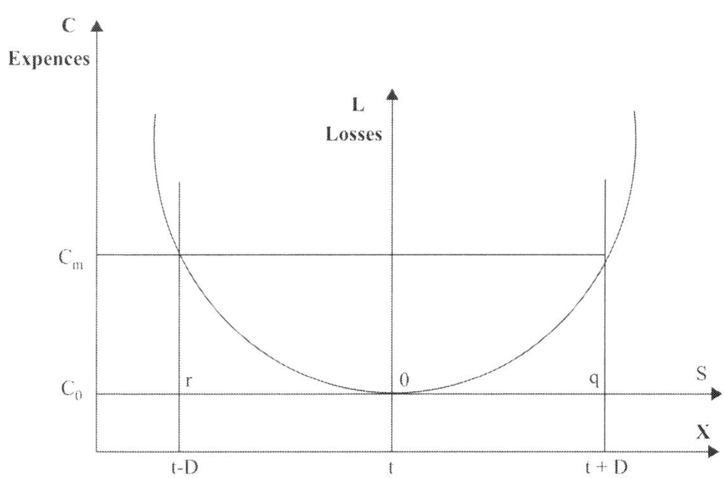

Figure II.11.5. Taguchi loss function.

As it can be seen in Figure II.11.5, the expenses for the product are minimal (C_0) with a value of the parameter equal to the target value *t*. The correlation between the expenses *L* and the deviation from the target value *S* has the following form:

$$L(x) = k \cdot S^2, \tag{II.11.23}$$

where *k* is the proportionality coefficient.

The interpretation of the Taguchi function shows that articles with parameters within the limits of tolerance are not all with the same quality. This is confirmed by the fact that the companies that work with a bigger reserve of technological accuracy (dispersion lower than the tolerance) with the same tolerance as their competition are the ones with higher quality of the production.

APPLICATION OF THE FUNCTION

1. Determination of the tolerance of functional parameters:
   In some cases, determination of the tolerance of the article's parameters is done on the basis of well-grounded methods as the conditions of exploitation, purpose, used material, and other requirements to the articles are taken in consideration. Unfortunately, in most cases there are no objective criteria for their choice and they are chosen by the constructor on the grounds of his practical experience and intuition. Often the size of the tolerance is determined also by the resources of the production equipment. The Taguchi approach allows for the tolerance to be determined depending on the loss of the customer and the producer.

Example

Determination of the optimal interval between standard sizes. It has been found that the tolerance limits of maximal admissible discomfort for an average customer are $r = -2.5$ cm and $q = 6$ cm. Above these deviations, the customer is willing to pay additionally 80 monetary units.

According to equation (II.11.23) two coefficients k can be determined that correspond to the lower r and the upper q tolerance limit:

$$k_1 = \frac{80}{2.5^2} = 12.8 \quad \text{and} \quad k_2 = \frac{80}{6^2} = 2.22.$$

If the interval between the standard sizes is 10 cm and a specific customer has a size which is bigger by 3 cm than a given standard size, he has to choose between an article with 3 cm tightness (buying the smaller size) or with 7 cm looseness (buying the bigger size). The loss for the customer in this case will be:

$$L_1(-3) = k_1(-3)^2 = 115.2 \text{ units and}$$

$$L_2(+7) = k_2(7)^2 = 108.78 \text{ units.}$$

As this loss is higher than the additional expenses for a special order (80 units), this customer will prefer a special workmanship.

The example shows that in the given case the interval T between the standard sizes is not optimal and has to be decreased. This however will lead to an increase of the expenses for the producer who will have to produce a higher number of standard sizes. The losses in the distribution network will also increase as more warehouse reserve and area will be necessary for them. If the loss for the distributor is 10 units per article, then they are lower than the production loss (80 units). In this case, the loss of the distributor can be used as a criterion for determination of the optimal interval between standard sizes. The loss of the customer and of the distributor will become equal as the optimal expenses will be $L_0 = 10$ units.

(*Continued*)

With a postulation that the value distribution within the tolerance is even (equally probable), then the mean value of loss can be determined by integration of equation (II.11.23) in the interval T:

$$L_{mean} = \frac{1}{T}\left(k_1 \int_{-r}^{0} S^2 dx + k_2 \int_{0}^{q} S^2 dx\right). \qquad \text{(II.11.24)}$$

Placing $L_{mean} = L_0 = 10$ units and replacing with the values k_1 and k_2, we get:

$$10 = \frac{1}{T}\left(\frac{12.8r^3}{3} + 2.22q^3\right).$$

Taking into consideration that the loss for the two limit tolerances is equal, from formula (II.11.23) the result is

$$k_1 r^2 = k_2 q^2. \qquad \text{(II.11.25)}$$

The optimal values are obtained by solving equations (II.11.24) and (II.11.25) together and by taking into consideration that $T = q + r$:

$q = 3.67$ cm; $r = 1.53$ cm; $T = 5.20$ cm.

The examined example shows that the methods can be used for determination of the optimal tolerance of a given functional parameter though analysis of the losses in the producer–distributor–consumer chain. It can also be acted analogically within a given production if as a "producer" we consider the executer of the initial operations, as a "distributor," the executer of the intermediate operations and as a "consumer" the executer of the finishing operations. For facility's sake we can agree that the tolerance is situated symmetrically toward the nominal value of the parameter. Even in cases of nonsymmetrical distribution, the tolerance can be seen as symmetrical towards the mean value.

Let us presume that the tolerance limits q_k of a given parameter have a critical value, that is, if these limits are exceeded, the functional characteristics of the article will be disturbed. In this case, the customer will have a financial loss:

$$L_k = kq_k^2,$$

that is, the proportionality coefficient is

$$k = \frac{L_k}{q_k^2}. \qquad \text{(II.11.26)}$$

The losses for the customer include direct and indirect expenses and can be determined by accounting: the transportation expenses for return of the article, the consumed time, the caused production complications, and so on. In most cases the exact determination is impossible but an approximate evaluation with an error of about 20–30 % can be determined. When the bought article is returned to the distributor, he will realize losses for transportation, warehouse space, labor, and other expenses related with the reclamation to the producer. If the loss for the distributor is L_d, the value of the tolerance limit q_d can be determined so that the loss for the producer does not exceed L_d. As $L_d = kq_d^2$, then for q_d can be written:

$$q_d = q_k \sqrt{\frac{L_d}{L_k}}. \tag{II.11.27}$$

However, the distributor cannot affect the quality of the article. A correction can only be made by the producer whose expenses for fault correction and second control are:

$$L_{i'} = kq_{i'}^2.$$

After replacement of the coefficient k, the optimal tolerance limit q_n can be determined and respectively the optimal tolerance T_p for the producer ($T = 2q$):

$$T_p = T_k \sqrt{\frac{L_p}{L_k}}. \tag{II.11.28}$$

The correlation shows that the tolerance of the producer depends on the tolerance of the customer. However, the loss for the distributor has to be added to the loss for the customer too.

2. Relation of the loss function to process stability:
The coefficient of process stability according to formula (I.4.1) is

$$C_p = \frac{T}{6\sigma}.$$

The mean loss for a unit of an article due to dispersion within the tolerance can be expressed with the variance σ^2 according to formula (II.11.23):

$$L_{mean} = k\sigma^2.$$

As $\sigma = \dfrac{T}{6 \cdot C_p}$, then the loss is:

$$L_{mean} = k\left(\frac{T}{6 \cdot C_p}\right)^2. \tag{II.11.29}$$

Example

The mean loss in the production of an article with tolerance $T = 5.2$ cm and a proportionality coefficient $k = 50$ should be determined for stability coefficients $C_p = 1$ and $C_p = 1.33$.

With $C_p = 1$, the mean loss for a unit of an article is:

$$L_{mean} = 50\left(\frac{5.2}{6\times 1}\right)^2 = 38 \text{ units.}$$

With $C_p = 1.33$, the mean loss for a unit of an article is:

$$L_{mean} = 50\left(\frac{5.2}{6\times 1.33}\right)^2 = 33 \text{ units.}$$

The difference between the expenses for the two stability coefficients is 5 units which, with an annual production of 100,000 pieces, gives a saving of 500,000 units.

With the suggested methods, a comparison between two technological processes with different values of L and C_p can be made. The loss is inversely proportional of the square of the stability coefficient:

$$\frac{L_1}{L_2} = \left(\frac{C_{p2}}{C_{p1}}\right)^2. \qquad \text{(II.11.30)}$$

The correlations obtained allow for a determination to be made on the need for improvement of the technological process. If the expenses for its improvement exceed the saved losses which are determined through the loss function, then improvements must not be made.

3. Determination of the tolerances of affecting parameters:
   The quality features of the end product depend on the tolerances of the parameters of the input materials (e.g., length of fibers, linear density, sliver irregularity, deviations from the width of the made-up textile article). If L_s is the loss for the subsupplier from rejection of his products, then the tolerance of the subsupplier can be determined from formula (II.11.28) as:

$$T_s = \frac{T_p}{p\sqrt{L_s / L_p}}, \qquad \text{(II.11.31)}$$

where p is the proportionality coefficient accounting the influence of a given parameter on the characteristics of the end product. It is determined by the ratio:

$$p = \frac{S_y}{S_x}, \qquad \text{(II.11.32)}$$

where S_x and S_y are the deviations from the nominal of the input and the output parameter.

4. Determination of the tolerances of parameter that change in time:
   A number of parameters of the article change in time or with exploitation. If q is the period for which the change is determined (this usually is the period of the article), then the tolerance of the subsupplier is determined as:

$$T_s = \frac{T_p}{p\theta\sqrt{3L_s / L_p}}. \tag{II.11.33}$$

   After a determination of the tolerances of all parameters affecting the characteristics of the end product $T_{c,i}$, the loss functions for each parameter $L_{c,i}$ are determined. The total loss due to deviations from the nominal values is

$$L = \sum_{i=1}^{n} \frac{L_{c,i}}{\left(\frac{T_{c,i}}{2}\right)^2}\sigma^2. \tag{II.11.34}$$

   The analysis of this function allows for a comparison to be made between articles from one and the same kind so that their quality is determined objectively.
5. Technical and economic ground and analysis of projects for technological, control, and other equipment:
   When the choice of machines, reference instruments, and so on is made, grounding has to be made not only for the main technical parameters, but also an economic grounding proving their profitability for the projected period of use. Taguchi suggests that the dispersion of the parameters of the ready production is used as a criterion of quality and the choice is made after an evaluation of the loss function. The total expenses for quality improvement are:

$$L = \frac{C}{N} + \frac{L_p}{\left(\frac{T_p}{2}\right)^2}\sigma^2, \tag{II.11.35}$$

   where C is the annual expenses for equipment,
   N is the number of articles produced per year.
   In this way, the expenses for quality granting can be determined even before the purchase of equipment. When a comparison is made with the existing condition regarding the quality and the expenses, the relation between the dispersion, the tolerances, and the stability coefficient of the existing process are used.
   For the output condition they are:

$$L = \frac{C_o}{N} + \frac{L_n}{\left(3C_p\right)^2}, \tag{II.11.36}$$

   where C_o is the existing annual expenses. The positive difference between formulae (II.11.35) and (II.11.36) is the profit from quality improvement.
   The presented system of methods with the use of the loss function offers the possibility of comparative analysis, an evaluation of the quality of technological and control resources, and well-grounded choice in consideration of technical and economical criteria.

APPENDIX 1

STUDENT'S *T*-DISTRIBUTION

as a function of the probability α and the degrees of freedom f

f \ α	0.20	0.10	0.05	0.025	0.01	0.005	0.001
1	3.078	6.314	12.706	25.452	63.656	127.321	636.578
2	1.886	2.920	4.303	6.205	9.925	14.089	31.600
3	1.638	2.353	3.182	4.177	5.841	7.453	12.924
4	1.533	2.132	2.776	3.495	4.604	5.598	8.610
5	1.476	2.015	2.571	3.163	4.032	4.773	6.869
6	1.440	1.943	2.447	2.969	3.707	4.317	5.959
7	1.415	1.895	2.365	2.841	3.499	4.029	5.408
8	1.397	1.860	2.306	2.752	3.355	3.833	5.041
9	1.383	1.833	2.262	2.685	3.250	3.690	4.781
10	1.372	1.812	2.228	2.634	3.169	3.581	4.587
11	1.363	1.796	2.201	2.593	3.106	3.497	4.437
12	1.356	1.782	2.179	2.560	3.055	3.428	4.318
13	1.350	1.771	2.160	2.533	3.012	3.372	4.221
14	1.345	1.761	2.145	2.510	2.977	3.326	4.140
15	1.341	1.753	2.131	2.490	2.947	3.286	4.073
16	1.337	1.746	2.120	2.473	2.921	3.252	4.015
17	1.333	1.740	2.110	2.458	2.898	3.222	3.965
18	1.330	1.734	2.101	2.445	2.878	3.197	3.922
19	1.328	1.729	2.093	2.433	2.861	3.174	3.883
20	1.325	1.725	2.086	2.423	2.845	3.153	3.850
21	1.323	1.721	2.080	2.414	2.831	3.135	3.819

(*Continued*)

(*Continued*)

f \ α	**0.20**	**0.10**	**0.05**	**0.025**	**0.01**	**0.005**	**0.001**
22	1.321	1.717	2.074	2.405	2.819	3.119	3.792
23	1.319	1.714	2.069	2.398	2.807	3.104	3.768
24	1.318	1.711	2.064	2.391	2.797	3.091	3.745
25	1.316	1.708	2.060	2.385	2.787	3.078	3.725
26	1.315	1.706	2.056	2.379	2.779	3.067	3.707
27	1.314	1.703	2.052	2.373	2.771	3.057	3.689
28	1.313	1.701	2.048	2.368	2.763	3.047	3.674
29	1.311	1.699	2.045	2.364	2.756	3.038	3.660
30	1.310	1.697	2.042	2.360	2.750	3.030	3.646
40	1.303	1.684	2.021	2.329	2.704	2.971	3.551
50	1.299	1.676	2.009	2.311	2.678	2.937	3.496
100	1.290	1.660	1.984	2.276	2.626	2.871	3.390
150	1.287	1.655	1.976	2.264	2.609	2.849	3.357
200	1.286	1.653	1.972	2.258	2.601	2.838	3.340
∞	1.282	1.645	1.960	2.241	2.576	2.807	3.290

APPENDIX 2

CHI-SQUARED χ^2-DISTRIBUTION

as a function of the probability α and the degrees of freedom f

f \ α	0.99	0.975	0.95	0.05	0.025	0.01
1	0.0002	0.001	0.004	3.841	5.024	6.635
2	0.020	0.051	0.103	5.991	7.378	9.210
3	0.115	0.216	0.352	7.815	9.348	11.345
4	0.297	0.484	0.711	9.488	11.143	13.277
5	0.554	0.831	1.145	11.070	12.832	15.086
6	0.872	1.237	1.635	12.592	14.449	16.812
7	1.239	1.690	2.167	14.067	16.013	18.475
8	1.647	2.180	2.733	15.507	17.535	20.090
9	2.088	2.700	3.325	16.919	19.023	21.666
10	2.558	3.247	3.940	18.307	20.483	23.209
11	3.053	3.816	4.575	19.675	21.920	24.725
12	3.571	4.404	5.226	21.026	23.337	26.217
13	4.107	5.009	5.892	22.362	24.736	27.688
14	4.660	5.629	6.571	23.685	26.119	29.141
15	5.229	6.262	7.261	24.996	27.488	30.578
16	5.812	6.908	7.962	26.296	28.845	32.000
17	6.408	7.564	8.672	27.587	30.191	33.409
18	7.015	8.231	9.390	28.869	31.526	34.805
19	7.633	8.907	10.117	30.144	32.852	36.191
20	8.260	9.591	10.851	31.410	34.170	37.566
21	8.897	10.283	11.591	32.671	35.479	38.932

(Continued)

(*Continued*)

f \ α	**0.99**	**0.975**	**0.95**	**0.05**	**0.025**	**0.01**
22	9.542	10.982	12.338	33.924	36.781	40.289
23	10.196	11.689	13.091	35.172	38.076	41.638
24	10.856	12.401	13.848	36.415	39.364	42.980
25	11.524	13.120	14.611	37.652	40.646	44.314
26	12.198	13.844	15.379	38.885	41.923	45.642
27	12.878	14.573	16.151	40.113	43.195	46.963
28	13.565	15.308	16.928	41.337	44.461	48.278
29	14.256	16.047	17.708	42.557	45.722	49.588
30	14.953	16.791	18.493	43.773	46.979	50.892
40	22.164	24.433	26.509	55.758	59.342	63.691
50	29.707	32.357	34.764	67.505	71.420	76.154
100	70.065	74.222	77.929	124.342	129.561	135.807
150	112.668	117.985	122.692	179.581	185.800	193.207
200	156.432	162.728	168.279	233.994	241.058	249.445

APPENDIX 3

FISHER'S *F*-DISTRIBUTION

for probability $\alpha = 0.05$

f_2 \ f_1	**1**	**2**	**3**	**4**	**5**	**6**	**7**	**8**	**9**	**10**	**15**	**20**	**30**	∞
1	161.4	199.5	215.7	224.6	230.2	234.0	236.8	238.9	240.5	241.9	245.9	248.0	250.1	254.3
2	18.51	19.00	19.16	19.25	19.30	19.33	19.35	19.37	19.38	19.40	19.43	19.45	19.46	19.50
3	10.13	9.55	9.28	9.12	9.01	8.94	8.89	8.85	8.81	8.79	8.70	8.66	8.62	8.53
4	7.71	6.94	6.59	6.39	6.26	6.16	6.09	6.04	6.00	5.96	5.86	5.80	5.75	5.63
5	6.61	5.79	5.41	5.19	5.05	4.95	4.88	4.82	4.77	4.74	4.62	4.56	4.50	4.36
6	5.99	5.14	4.76	4.53	4.39	4.28	4.21	4.15	4.10	4.06	3.94	3.87	3.81	3.67
7	5.59	4.74	4.35	4.12	3.97	3.87	3.79	3.73	3.68	3.64	3.51	3.44	3.38	3.23
8	5.32	4.46	4.07	3.84	3.69	3.58	3.50	3.44	3.39	3.35	3.22	3.15	3.08	2.93
9	5.12	4.26	3.86	3.63	3.48	3.37	3.29	3.23	3.18	3.14	3.01	2.94	2.86	2.71
10	4.96	4.10	3.71	3.48	3.33	3.22	3.14	3.07	3.02	2.98	2.85	2.77	2.70	2.54
11	4.84	3.98	3.59	3.36	3.20	3.09	3.01	2.95	2.90	2.85	2.72	2.65	2.57	2.40
12	4.75	3.89	3.49	3.26	3.11	3.00	2.91	2.85	2.80	2.75	2.62	2.54	2.47	2.30
13	4.67	3.81	3.41	3.18	3.03	2.92	2.83	2.77	2.71	2.67	2.53	2.46	2.38	2.21
14	4.60	3.74	3.34	3.11	2.96	2.85	2.76	2.70	2.65	2.60	2.46	2.39	2.31	2.13
15	4.54	3.68	3.29	3.06	2.90	2.79	2.71	2.64	2.59	2.54	2.40	2.33	2.25	2.07
16	4.49	3.63	3.24	3.01	2.85	2.74	2.66	2.59	2.54	2.49	2.35	2.28	2.19	2.01
17	4.45	3.59	3.20	2.96	2.81	2.70	2.61	2.55	2.49	2.45	2.31	2.23	2.15	1.96
18	4.41	3.55	3.16	2.93	2.77	2.66	2.58	2.51	2.46	2.41	2.27	2.19	2.11	1.92
19	4.38	3.52	3.13	2.90	2.74	2.63	2.54	2.48	2.42	2.38	2.23	2.16	2.07	1.88
20	4.35	3.49	3.10	2.87	2.71	2.60	2.51	2.45	2.39	2.35	2.20	2.12	2.04	1.84
21	4.32	3.47	3.07	2.84	2.68	2.57	2.49	2.42	2.37	2.32	2.18	2.10	2.01	1.81

(*Continued*)

22	4.30	3.44	3.05	2.82	2.66	2.55	2.46	2.40	2.34	2.30	2.15	2.07	1.98	1.78
23	4.28	3.42	3.03	2.80	2.64	2.53	2.44	2.37	2.32	2.27	2.13	2.05	1.96	1.76
24	4.26	3.40	3.01	2.78	2.62	2.51	2.42	2.36	2.30	2.25	2.11	2.03	1.94	1.73
25	4.24	3.39	2.99	2.76	2.60	2.49	2.40	2.34	2.28	2.24	2.09	2.01	1.92	1.71
26	4.23	3.37	2.98	2.74	2.59	2.47	2.39	2.32	2.27	2.22	2.07	1.99	1.90	1.69
27	4.21	3.35	2.96	2.73	2.57	2.46	2.37	2.31	2.25	2.20	2.06	1.97	1.88	1.67
28	4.20	3.34	2.95	2.71	2.56	2.45	2.36	2.29	2.24	2.19	2.04	1.96	1.87	1.65
29	4.18	3.33	2.93	2.70	2.55	2.43	2.35	2.28	2.22	2.18	2.03	1.94	1.85	1.64
30	4.17	3.32	2.92	2.69	2.53	2.42	2.33	2.27	2.21	2.16	2.01	1.93	1.84	1.62
40	4.08	3.23	2.84	2.61	2.45	2.34	2.25	2.18	2.12	2.08	1.92	1.84	1.74	1.51
50	4.03	3.18	2.79	2.56	2.40	2.29	2.20	2.13	2.07	2.03	1.87	1.78	1.69	1.44
100	3.94	3.09	2.70	2.46	2.31	2.19	2.10	2.03	1.97	1.93	1.77	1.68	1.57	1.28
150	3.90	3.06	2.66	2.43	2.27	2.16	2.07	2.00	1.94	1.89	1.73	1.64	1.54	1.22
200	3.89	3.04	2.65	2.42	2.26	2.14	2.06	1.98	1.93	1.88	1.72	1.62	1.52	1.19
∞	3.84	3.00	2.60	2.37	2.21	2.10	2.01	1.94	1.88	1.83	1.67	1.57	1.46	1.00

APPENDIX 4

FISHER'S *F*-DISTRIBUTION

for probability $\alpha = 0.01$

f_2 \ f_1	**1**	**2**	**3**	**4**	**5**	**6**	**7**	**8**	**9**	**10**	**15**	**20**	**30**	∞
1	4052	4999	5404	5624	5764	5859	5928	5981	6022	6056	6157	6209	6240	6260
2	98.50	99.00	99.16	99.25	99.30	99.33	99.36	99.38	99.39	99.40	99.43	99.45	99.46	99.47
3	34.12	30.82	29.46	28.71	28.24	27.91	27.67	27.49	27.34	27.23	26.87	26.69	26.58	26.50
4	21.20	18.00	16.69	15.98	15.52	15.21	14.98	14.80	14.66	14.55	14.20	14.02	13.91	13.84
5	16.26	13.27	12.06	11.39	10.97	10.67	10.46	10.29	10.16	10.05	9.72	9.55	9.45	9.38
6	13.75	10.92	9.78	9.15	8.75	8.47	8.26	8.10	7.98	7.87	7.56	7.40	7.30	7.23
7	12.25	9.55	8.45	7.85	7.46	7.19	6.99	6.84	6.72	6.62	6.31	6.16	6.06	5.99
8	11.26	8.65	7.59	7.01	6.63	6.37	6.18	6.03	5.91	5.81	5.52	5.36	5.26	5.20
9	10.56	8.02	6.99	6.42	6.06	5.80	5.61	5.47	5.35	5.26	4.96	4.81	4.71	4.65
10	10.04	7.56	6.55	5.99	5.64	5.39	5.20	5.06	4.94	4.85	4.56	4.41	4.31	4.25
11	9.65	7.21	6.22	5.67	5.32	5.07	4.89	4.74	4.63	4.54	4.25	4.10	4.01	3.94
12	9.33	6.93	5.95	5.41	5.06	4.82	4.64	4.50	4.39	4.30	4.01	3.86	3.76	3.70
13	9.07	6.70	5.74	5.21	4.86	4.62	4.44	4.30	4.19	4.10	3.82	3.66	3.57	3.51
14	8.86	6.51	5.56	5.04	4.69	4.46	4.28	4.14	4.03	3.94	3.66	3.51	3.41	3.35
15	8.68	6.36	5.42	4.89	4.56	4.32	4.14	4.00	3.89	3.80	3.52	3.37	3.28	3.21
16	8.53	6.23	5.29	4.77	4.44	4.20	4.03	3.89	3.78	3.69	3.41	3.26	3.16	3.10
17	8.40	6.11	5.19	4.67	4.34	4.10	3.93	3.79	3.68	3.59	3.31	3.16	3.07	3.00
18	8.29	6.01	5.09	4.58	4.25	4.01	3.84	3.71	3.60	3.51	3.23	3.08	2.98	2.92
19	8.18	5.93	5.01	4.50	4.17	3.94	3.77	3.63	3.52	3.43	3.15	3.00	2.91	2.84
20	8.10	5.85	4.94	4.43	4.10	3.87	3.70	3.56	3.46	3.37	3.09	2.94	2.84	2.78
21	8.02	5.78	4.87	4.37	4.04	3.81	3.64	3.51	3.40	3.31	3.03	2.88	2.79	2.72

(Continued)

22	7.95	5.72	4.82	4.31	3.99	3.76	3.59	3.45	3.35	3.26	2.98	2.83	2.73	2.67
23	7.88	5.66	4.76	4.26	3.94	3.71	3.54	3.41	3.30	3.21	2.93	2.78	2.69	2.62
24	7.82	5.61	4.72	4.22	3.90	3.67	3.50	3.36	3.26	3.17	2.89	2.74	2.64	2.58
25	7.77	5.57	4.68	4.18	3.85	3.63	3.46	3.32	3.22	3.13	2.85	2.70	2.60	2.54
26	7.72	5.53	4.64	4.14	3.82	3.59	3.42	3.29	3.18	3.09	2.81	2.66	2.57	2.50
27	7.68	5.49	4.60	4.11	3.78	3.56	3.39	3.26	3.15	3.06	2.78	2.63	2.54	2.47
28	7.64	5.45	4.57	4.07	3.75	3.53	3.36	3.23	3.12	3.03	2.75	2.60	2.51	2.44
29	7.60	5.42	4.54	4.04	3.73	3.50	3.33	3.20	3.09	3.00	2.73	2.57	2.48	2.41
30	7.56	5.39	4.51	4.02	3.70	3.47	3.30	3.17	3.07	2.98	2.70	2.55	2.45	2.39
40	7.31	5.18	4.31	3.83	3.51	3.29	3.12	2.99	2.89	2.80	2.52	2.37	2.27	2.20
50	7.17	5.06	4.20	3.72	3.41	3.19	3.02	2.89	2.78	2.70	2.42	2.27	2.17	2.10
100	6.90	4.82	3.98	3.51	3.21	2.99	2.82	2.69	2.59	2.50	2.22	2.07	1.97	1.89
150	6.81	4.75	3.91	3.45	3.14	2.92	2.76	2.63	2.53	2.44	2.16	2.00	1.90	1.83
200	6.76	4.71	3.88	3.41	3.11	2.89	2.73	2.60	2.50	2.41	2.13	1.97	1.87	1.79
∞	6.63	4.61	3.78	3.32	3.02	2.80	2.64	2.51	2.41	2.32	2.04	1.88	1.77	1.70

APPENDIX 5

COCHRAN'S CRITICAL VALUES

f_2 \ f_1	1	2	3	4	5	6	7	8	9	10	15	20	30	∞
Probability $\alpha = 0.05$														
2	0.999	0.975	0.939	0.906	0.877	0.853	0.833	0.816	0.801	0.788	0.734	0.660	0.581	0.500
3	0.967	0.871	0.798	0.746	0.707	0.677	0.653	0.633	0.617	0.603	0.547	0.475	0.403	0.333
4	0.907	0.768	0.684	0.629	0.590	0.560	0.537	0.518	0.502	0.488	0.437	0.372	0.309	0.250
5	0.841	0.684	0.598	0.544	0.507	0.478	0.456	0.439	0.424	0.412	0.365	0.307	0.251	0.200
6	0.781	0.616	0.532	0.480	0.445	0.418	0.398	0.382	0.368	0.357	0.314	0.261	0.212	0.167
7	0.727	0.561	0.480	0.431	0.397	0.373	0.354	0.338	0.326	0.315	0.276	0.228	0.183	0.143
8	0.680	0.516	0.438	0.391	0.360	0.336	0.319	0.304	0.293	0.283	0.246	0.202	0.162	0.125
9	0.639	0.478	0.403	0.358	0.329	0.307	0.290	0.277	0.266	0.257	0.223	0.182	0.145	0.111
10	0.602	0.445	0.373	0.331	0.303	0.282	0.267	0.254	0.244	0.235	0.203	0.166	0.131	0.100
12	0.541	0.392	0.326	0.288	0.262	0.244	0.230	0.219	0.210	0.202	0.174	0.140	0.110	0.083
15	0.471	0.335	0.276	0.242	0.220	0.203	0.191	0.182	0.174	0.167	0.143	0.114	0.089	0.067
20	0.389	0.271	0.221	0.192	0.174	0.160	0.150	0.142	0.136	0.130	0.111	0.088	0.068	0.050
24	0.343	0.235	0.191	0.166	0.149	0.137	0.129	0.122	0.116	0.111	0.094	0.074	0.057	0.042
30	0.293	0.198	0.159	0.138	0.124	0.114	0.106	0.100	0.096	0.092	0.077	0.060	0.046	0.033
40	0.237	0.158	0.126	0.108	0.097	0.089	0.083	0.078	0.075	0.071	0.060	0.046	0.035	0.025
60	0.174	0.113	0.090	0.077	0.068	0.062	0.058	0.055	0.052	0.050	0.041	0.032	0.023	0.017
120	0.100	0.063	0.050	0.042	0.037	0.034	0.031	0.029	0.028	0.027	0.022	0.017	0.012	0.008
∞	0	0	0	0	0	0	0	0	0	0	0	0	0	0

(Continued)

							Probability α = 0.01							
2	1.000	0.995	0.979	0.959	0.937	0.917	0.900	0.882	0.867	0.854	0.795	0.707	0.606	0.500
3	0.993	0.942	0.883	0.834	0.793	0.761	0.734	0.711	0.691	0.674	0.606	0.515	0.423	0.333
4	0.968	0.864	0.781	0.721	0.676	0.641	0.613	0.590	0.570	0.554	0.488	0.406	0.325	0.250
5	0.928	0.789	0.696	0.633	0.588	0.553	0.526	0.504	0.485	0.470	0.409	0.335	0.264	0.200
6	0.883	0.722	0.626	0.564	0.520	0.487	0.461	0.440	0.423	0.408	0.353	0.286	0.223	0.167
							Probability α = 0.01							
7	0.838	0.664	0.569	0.508	0.466	0.435	0.411	0.391	0.375	0.362	0.311	0.249	0.193	0.143
8	0.795	0.615	0.521	0.463	0.423	0.393	0.370	0.352	0.337	0.325	0.278	0.221	0.170	0.125
9	0.754	0.573	0.481	0.425	0.387	0.359	0.338	0.321	0.307	0.295	0.251	0.199	0.152	0.111
10	0.718	0.536	0.447	0.393	0.357	0.331	0.311	0.295	0.281	0.270	0.230	0.181	0.138	0.100
12	0.653	0.475	0.392	0.343	0.310	0.286	0.268	0.254	0.242	0.232	0.196	0.154	0.116	0.083
15	0.575	0.407	0.332	0.288	0.259	0.239	0.223	0.210	0.200	0.192	0.161	0.125	0.093	0.067
20	0.480	0.330	0.265	0.229	0.205	0.188	0.175	0.165	0.157	0.150	0.125	0.960	0.071	0.050
24	0.425	0.287	0.230	0.197	0.176	0.161	0.150	0.141	0.134	0.128	0.106	0.081	0.060	0.042
30	0.363	0.241	0.191	0.164	0.145	0.133	0.123	0.116	0.110	0.105	0.087	0.066	0.048	0.033
40	0.294	0.192	0.151	0.128	0.114	0.103	0.096	0.090	0.085	0.082	0.067	0.050	0.036	0.025
60	0.215	0.137	0.107	0.090	0.080	0.072	0.067	0.063	0.059	0.057	0.046	0.034	0.025	0.017
120	0.123	0.076	0.059	0.049	0.043	0.039	0.036	0.033	0.032	0.030	0.024	0.018	0.013	0.008
∞	0	0	0	0	0	0	0	0	0	0	0	0	0	0

APPENDIX 6

DENSITY OF NORMAL DISTRIBUTION $N(0, 1)$

x	0	1	2	3	4	5	6	7	8	9
0.0	0.3989	0.3989	0.3989	0.3988	0.3986	0.3984	0.3982	0.3980	0.3977	0.3973
0.1	0.3970	0.3965	0.3961	0.3956	0.3951	0.3945	0.3939	0.3932	0.3925	0.3918
0.2	0.3910	0.3902	0.3894	0.3885	0.3876	0.3867	0.3857	0.3847	0.3836	0.3825
0.3	0.3814	0.3802	0.3790	0.3778	0.3765	0.3752	0.3739	0.3725	0.3712	0.3697
0.4	0.3683	0.3668	0.3653	0.3637	0.3621	0.3605	0.3589	0.3572	0.3555	0.3538
0.5	0.3521	0.3503	0.3485	0.3467	0.3448	0.3429	0.3410	0.3391	0.3372	0.3352
0.6	0.3332	0.3312	0.3292	0.3271	0.3251	0.3230	0.3209	0.3187	0.3166	0.3144
0.7	0.3123	0.3101	0.3079	0.3056	0.3034	0.3011	0.2989	0.2966	0.2943	0.2920
0.8	0.2897	0.2874	0.2850	0.2827	0.2803	0.2780	0.2756	0.2732	0.2709	0.2685
0.9	0.2661	0.2637	0.2613	0.2589	0.2565	0.2541	0.2516	0.2492	0.2468	0.2444
1.0	0.2420	0.2396	0.2371	0.2347	0.2323	0.2299	0.2275	0.2251	0.2227	0.2203
1.1	0.2179	0.2155	0.2131	0.2107	0.2083	0.2059	0.2036	0.2012	0.1989	0.1965
1.2	0.1942	0.1919	0.1895	0.1872	0.1849	0.1826	0.1804	0.1781	0.1758	0.1736
1.3	0.1714	0.1691	0.1669	0.1647	0.1626	0.1604	0.1582	0.1561	0.1539	0.1518
1.4	0.1497	0.1476	0.1456	0.1435	0.1415	0.1394	0.1374	0.1354	0.1334	0.1315
1.5	0.1295	0.1276	0.1257	0.1238	0.1219	0.1200	0.1182	0.1163	0.1145	0.1127
1.6	0.1109	0.1092	0.1074	0.1057	0.1040	0.1023	0.1006	0.0989	0.0973	0.0957
1.7	0.0940	0.0925	0.0909	0.0893	0.0878	0.0863	0.0848	0.0833	0.0818	0.0804
1.8	0.0790	0.0775	0.0761	0.0748	0.0734	0.0721	0.0707	0.0694	0.0681	0.0669
1.9	0.0656	0.0644	0.0632	0.0620	0.0608	0.0596	0.0584	0.0573	0.0562	0.0551
2.0	0.0540	0.0529	0.0519	0.0508	0.0498	0.0488	0.0478	0.0468	0.0459	0.0449
2.1	0.0440	0.0431	0.0422	0.0413	0.0404	0.0396	0.0387	0.0379	0.0371	0.0363
2.2	0.0355	0.0347	0.0339	0.0332	0.0325	0.0317	0.0310	0.0303	0.0297	0.0290

(*Continued*)

(Continued)

x	0	1	2	3	4	5	6	7	8	9
2.3	0.0283	0.0277	0.0270	0.0264	0.0258	0.0252	0.0246	0.0241	0.0235	0.0229
2.4	0.0224	0.0219	0.0213	0.0208	0.0203	0.0198	0.0194	0.0189	0.0184	0.0180
2.5	0.0175	0.0171	0.0167	0.0163	0.0158	0.0154	0.0151	0.0147	0.0143	0.0139
2.6	0.0136	0.0132	0.0129	0.0126	0.0122	0.0119	0.0116	0.0113	0.0110	0.0107
2.7	0.0104	0.0101	0.0099	0.0096	0.0093	0.0091	0.0088	0.0086	0.0084	0.0081
2.8	0.0079	0.0077	0.0075	0.0073	0.0071	0.0069	0.0067	0.0065	0.0063	0.0061
2.9	0.0060	0.0058	0.0056	0.0055	0.0053	0.0051	0.0050	0.0048	0.0047	0.0046
3.0	0.0044	0.0043	0.0042	0.0040	0.0039	0.0038	0.0037	0.0036	0.0035	0.0034
3.1	0.0033	0.0032	0.0031	0.0030	0.0029	0.0028	0.0027	0.0026	0.0025	0.0025
3.2	0.0024	0.0023	0.0022	0.0022	0.0021	0.0020	0.0020	0.0019	0.0018	0.0018
3.3	0.0017	0.0017	0.0016	0.0016	0.0015	0.0015	0.0014	0.0014	0.0013	0.0013
3.4	0.0012	0.0012	0.0012	0.0011	0.0011	0.0010	0.0010	0.0010	0.0009	0.0009
3.5	0.0009	0.0008	0.0008	0.0008	0.0008	0.0007	0.0007	0.0007	0.0007	0.0006

The probabilities of the negative arguments $P(-x) = P(x)$.

APPENDIX 7

PROBABILITY *P* FOR $S_d \leq S_{do}$

n = 4		*n* = 5		*n* = 6		*n* = 7		*n* = 8		*n* = 9		*n* = 10	
S_0	*p*	S_0	*p*	S_0	*pw*	S_0	*p*	S_0	*p*	S_0	*p*	S_0	*p*
10	0.542	20	0.525	34	0.500	50	0.420	70	0.352	120	0.509	150	0.405
8	0.458	18	0.475	30	0.401	40	0.278	60	0.250	100	0.339	120	0.224
6	0.375	16	0.392	26	0.320	32	0.177	50	0.163	90	0.260	100	0.132
4	0.208	14	0.342	22	0249	28	0.133	40	0.098	80	0.193	80	0.067
2	0.167	12	0.258	20	0.210	24	0.100	32	0.057	70	0.0135	60	0.027
0	0.042	10	0 225	18	0.178	20	0.069	28	0.042	60	0.089	56	0.022
		8	0.175	16	0.149	18	0.055	24	0.029	50	0.054	52	0.017
		6	0.117	14	0.121	16	0.044	20	0.018	40	0029	48	0.013
		4	0 067	12	0.088	14	0.033	18	0.014	36	0.022	44	0.010
		2	0.042	10	0.068	12	0.024	16	0.011	32	0.016	40	0.007
		0	0.0083	8	0 051	10	0.017	14	0.007	28	0.011	36	0.005
				6	0.029	8	0.012	12	0.005	24	0.006	32	0.003
				4	0.017	6	0.006	10	0.003	20	0.004	28	0.002
				2	0.008	4	0.003	8	0.002	16	0C02	24	0.001
				0	0 001	2	0.001	6	0.001	12	0.001	20	0.001

APPENDIX 8

RANDOM NUMBERS

3820	1007	5964	8990	8845	9584	0145	4074	8632	1386	2450
0324	1641	2196	0171	2850	3431	5536	3573	3718	3556	9102
4261	3039	9756	8066	9911	2562	9516	0534	7050	8164	9724
3002	7501	3514	7756	0743	1984	0641	3583	4870	5112	3734
0407	2307	0050	9261	1003	2567	7756	6796	8090	7243	0850
7561	6265	1736	4048	5523	7114	5551	1811	9702	6869	5287
8056	2622	1779	8667	1148	0595	7615	7383	9862	9255	9038
5007	6749	4898	1458	0380	7962	6715	7316	5845	1522	8921
2005	2058	3339	3251	3002	8021	6960	2715	9040	0391	7090
5166	2565	2913	8021	7889	6759	7553	9484	6193	7220	9680
8503	5570	8730	4410	2177	8589	2803	7032	7073	3758	3297
9768	2855	5343	4074	9976	8946	8107	9085	5745	7060	4014
8973	3863	0958	7776	7835	6657	6568	2584	7651	7002	8587
6785	9287	0425	5181	9120	9542	5943	5576	9681	4830	2556
4960	8506	6680	9268	4517	1681	0619	0052	5410	6175	4929
6018	9300	5339	1320	0823	5759	8291	0657	2709	6996	4142
4350	3301	2111	7404	5234	8967	6034	5227	5897	5849	4970
5930	5589	7740	2308	7311	5867	5455	8073	9642	0954	1085
8830	1899	0160	1845	5800	6660	1623	1951	6778	5483	2945
1732	1853	8523	9480	2507	4317	5448	9674	7240	9509	5702
2598	1670	8837	8208	0419	8982	4208	1282	0301	2047	6820
4723	4785	5849	3624	8232	3105	9903	7715	7292	1633	8082
2323	3816	0902	9119	8519	5731	0142	6933	7006	8879	1690
2871	2316	0423	9153	1672	9388	1215	3412	0955	9440	5114
8355	9743	4757	1516	2845	8016	8084	6952	0686	0813	4428

(Continued)

(*Continued*)

2651	8102	2009	4542	4082	9355	0938	1748	4335	1443	0758
7192	3673	6600	0202	8784	0257	3024	1644	4599	5536	9586
2134	2276	7212	9003	0756	8337	9437	2524	5331	2037	7566
5185	1515	3822	7656	4961	8418	1553	7759	8927	1211	6541
5315	8428	8497	5418	2231	7185	6787	7669	1713	7246	9463
3500	5982	9654	0082	5066	4510	8387	8917	9489	0345	7892
7709	6857	4461	9514	6752	4875	4917	4790	0462	6711	5764
4329	7954	9067	9713	0950	7323	4146	2290	7701	9900	9113
3180	4059	1361	5298	3979	7387	1013	6981	8355	1938	6448
8353	4103	5921	3375	6950	9425	4368	1530	1784	7672	9447

Bibliography

Adler, U. *Vvedenie n planirovanie experimenta*. Moskva, 1969. (In Russian.)

Adler, U., and E. Markova. *Planirovanie experimenta v istoricheskom aspekte*. Moskva, 1970. (In Russian.)

Adler, U., et al. *Planirovanie experimenta pri poiske optimalnih uslovii*. Moskva: Nauka, 1976. (In Russian.)

Ahrens, H., and J. Läuter. *Mehrdimensionale Varianzanalyse*. Berlin: Akademie-Verlag, 1981.

Andonova, Sn. Algoritam za matematichesko modelirane na prozesa na vlagotoplinna obrabotka v shevnoto proisvodstvo. *Tekstil i obleklo* 8–9, 2004. (In Bulgarian.)

Andonova, Sn. Applying a statistical method for establishing some substantial factors of a technological processes. Paper presented at 10th International Research/Expert Conference TMT 2006, Barcelona-Lloret de Mar, Spain, September 11–15, 2006.

Andonova, Sn. *Uchebno-metodichno posobie po Prilozhenie na matematicheskite metodi v shevnite technologi*. Blagoevgrad: Universitetsko izdatelstvo na UZO "Neolfit Rilski," 2006.

Bellman, R. *Vvedenie v teorii matriz*. Moskva: Nauka, 1969. (In Russian.)

Box, G. E. P. and Wilson, K.B. (1951) On the Experimental Attainment of Optimum Conditions (with discussion). *Journal of the Royal Statistical Society* Series B 13(1):1–45

Boyarinova, A., and V. Kafarov. *Metodi optimisazii v himicheskoi technologii*. Moskva: Himia, 1969. (In Russian.)

Bozhanov, E., and I. Vuchkov. *Statisticheski metodi za modelirane i optimizirane na mnogofaktorni obekti*. Sofia: Technika, 1973. (In Bulgarian.)

Bozhanov, E., and I. Vuchkov. *Statisticheski reshenia v proizvodstvoto i nauchnite izsledvania*. Sofia: Technika, 1979. (In Bulgarian.)

Bronshtain, I., and Semediaev. *Spravochnik po matematike*. Moskva: Nauka, 1967. (In Russian.)

Damyanov, G. *Matematicheski metodi za planirane na experimenta pri izslednania v tekstilnata promishlenost*. Sofia: TU-Sofia, 1977. (In Bulgarian.)

Damyanov, G. Optimizirane na takachnia prozes s pomoshta na zentralno-komposizionen rotatabilen experiment. PhD thesis, Moskva, 1974. (In Russian.)

Damyanov, G. Prilojenie na zentralno-komposizionnia rotatabilen experiment za optimizirane na prozesa pri stanove MAV. *Tekstilna promishlenost*, 1976. (In Bulgarian.)

Damyanov, G., and K. Simonski. Izsledvane strukturite na takani s pomoshta na isotopi. *Tekstilna promishlenost*, 1974. (In Bulgarian.)

Danov, T. *Statisticheski metodi za upravlenie na kachestvoto*. Sofia: Technika, 1987. (In Bulgarian.)

Diukendzhiev, G., and R. Yordanov. *Kontrol i upravlenie na kachestvoto*. Sofia: Softrade, 2002. (In Bulgarian.)

Draper, N., and G. Smith. *Prikladnoi regressionnii analiz*. Moskva, 1973. (In Russian.)

Feodorov, V. *Teoria optimalnogo experimenta*. Moskva: Nauka, 1971. (In Russian.)

Georgiev, I., and D. Germanova-Krasteva. Nasoki v razvitieto na modeliraneto na technologichnite prozesi. *Tekstil i obleklo* 6–7, 1997. (In Bulgarian.)

Germanova-Krasteva, D. Simulaziono modelirane na pamuchni lenti. PhD thesis, Sofia, 2005. (In Bulgarian.)

Germanova-Krasteva, D. *Rakovodstvo za laboratorni uprazhnenia po Tekstilni izpitvania i analiz*. Sofia: TU-Sofia, 2007. (In Bulgarian.)

Germanova-Krasteva, D., and Ch. Petrov. Investigation on the seam's quality by sewing of light fabrics. *International Journal of Clothing Science and Technology* 20, 57–64, 2008. DOI: 10.1108/09556220810843539

Germanova-Krasteva, D., and G. Kandzhikova. Optimizirane na parametri, harakterizirashti opipa na havlieni takani. *Tekstil i obleklo* 7, 2009. (In Bulgarian.)

Germanova-Krasteva, D., and I. Georgiev. Cotton sliver simulation modelling. *International Textile Bulletin* 47, 58; 60–62, 2001.

Germanova-Krasteva, D., and I. Georgiev. Modell- und Simulationsstrecken von Baumwollbänder. *Melliand Textilberichte* 4, 2006. (In German.)

Germanova-Krasteva, D., and I. Georgiev. Simulation der Häkchenauflösung der Baumwollfasern beim Strecken. *Melliand Textilberichte* 1–2, 2007. (In German.)

Gmurman, V. E. *Teoria veroiatnostei i matematicheskaia statistika*. Visshaia shkola, 1977. (In Russian.)

Golikov, T. *Katalog planov vtorova poriadka*. Moskva: MGU, 1973. (In Russian.)

Gorskii, V., and U. Adler. *Planirovanie promishlenih experimentov*. Moskva, 1974. (In Russian.)

Graf, U., and H.-J. Henning, *Statistische Methoden bei textilen Untersuchungen*. Berlin/Göttingen/Heidelberg: Springer-Verlag, 1952. (In German.)

Han, G., and S. Shariro. *Statisticheskie modeli v inzhinernih zadach*. Moskva: Mir, 1969. (In Russian.)

Hicks, Ch., and K. Turner Jr. *Fundamental Concepts in the Design of Experiments*. USA: Oxford University Press, 1999.

Kamiske, G., and J.-P. Brauer. *Qualitätsmanagement von A bis Z*. München Wien: Karl Hanser Verlag, 1993. (In German.)

Kendall, M., and J. Gibbons. *Rank Correlation Methods*. A Charles Griffin Book, 1990.

Kevorkian, A., and G. Georgiev. *Prilojenie na matematicheskata statistika v tekstilnata promishlenost*. Sofia: Technika, 1969. (In Bulgarian.)

Kevorkian, A., et al. *Statisticheski control v tekstilnata i shevnata promishlenost*. Sofia: Technika, 1980. (In Bulgarian.)

Klemm, L., et al. *Statistische Kontrollmethoden in der Textilindustrie*. Leipzig: VEB Fachbuchverlag, 1968. (In German.)

Law, A., and W. Kelton. *Simulation Modeling and Analysis*. New York: McGraw-Hill, 1996.

Lukomskii, Ia. *Teoria korelazii i ieo primenie k analizu proizvodstva*. Moskva: Gostatizdat, 1961. (In Russian.)

Markova, E. *Nepolnoblochnie plani*. Moskva: MGU, 1970. (In Russian.)

Markova, E., and A. Lisenkov. *Planivovanie experimenta v usloviah neednorodnostei*. Moskva: Nauka, 1973. (In Russian.)

Mitova, B., and D. Germanova-Krasteva. *Rakovodstvo za laboratorni uprazhnenia po Tekstilni izpitvania*. Sofia: TU-Sofia, 1995. (In Bulgarian.)

Murphy, T., K. P. Norris, and L. H. C. Tippett. *Statistical Methods for Textile Technologists*. Manchester: Textile Institute, 1979.

Nalimov, V. *Teoria experimenta*. Moskva: Nauka, 1971. (In Russian.)

Nalimov, V., and N. Cherneva. *Statisticheskie metodi planirovania extremalnih experimentov*. Moskva: Nauka, 1965. (In Russian.)

Nikitina, E. *Planirovanie i analiz experimenta*. Moskva, 1976. (In Russian.)

Popov, N., and A. Panov. *Eksperimentalni izslednania i avtomatizazia v tekstilnoto i konfekzionoto proizvodstvo*. Sofia: Technika, 1980. (In Bulgarian.)

Sevostianov, A., and P. Sevostianov. *Optimisazia mechaniko-technologicheskih prozesov tekstilnoi promishlenosti*. Moskva: Legprombitizdat, 1991. (In Russian.)

Shtut, I. Izpolsovanie simplex planov pri proektirovanie smesei. *Tesksilnaia Promishlenost* 5, 1972. (In Russian.)

Simonenko, D. Planirovanie experimenta pri izsloedovania trikotazha. *Legkaia Promishlenost* 6. 1972. (In Russian.)

Stoyanov, S. *Optimisazia na technologichni procesi*. Sofia: Technika, 1993. (In Bulgarian.)

Tichomirov, V. *Matematicheskie metodi planirovania experimenta pri izuchenia netkanih materialov*. Moskva, 1968. (In Russian.)

Ventzel, E. *Teoria veroiatnostei*. Moskva: Nauka, 1969. (In Russian.)

Vinogradov, U. *Matematicheskaia statistika i eio primenenie v tekstilnoi promishlenosti*. Moskva, 1970. (In Russian.)

Vosnesenskii, B. *Statisticheskie metodi planirovania experimenta v technico-ikonomicheskii izsledovanii*. Moskva, 1974. (In Russian.)

Vuchkov, I. *Optimalno planirane na eksperimentalnite izsledvania*. Sofia: Technika, 1978. (In Bulgarian.)

Vuchkov, I., and S. Stoyanov. *Matematichesko modelirane i optimisazia na technologichni obekti*. Sofia: Technika, 1979. (In Bulgarian.)

Vuchkov, I., S. Stoyanov, and K. Velev. *Rakovodstvo za laboratorni uprazhnenia po matematichesko modelirane i optimisazia na technologichni obekti*. Sofia: Technika, 1986. (In Bulgarian.)

Yordanov, R., and G. Diukendzhiev. *Metodi za ozenka na kachestvoto*. Sofia: Softrade, 2002. (In Bulgarian.)

INDEX

CPSIA information can be obtained at www.ICGtesting.com
Printed in the USA
BVOW060003291012
304056BV00003B/3/P